Ebola Virus Disease

Marta Lado

Editor

Ebola Virus Disease

A Manual for EVD Management

 Springer

Editor
Marta Lado
King's Sierra Leone Partnership
Freetown, Sierra Leone

ISBN 978-3-319-94853-9   ISBN 978-3-319-94854-6   (eBook)
https://doi.org/10.1007/978-3-319-94854-6

Library of Congress Control Number: 2018965918

This Springer imprint is published by the registered company Springer Nature Switzerland AG
The registered company address is: Gewerbestrasse 11, 6330 Cham, Switzerland

Contents

Introduction to Viral Haemorrhagic Fevers

Colin S. Brown and Oliver Johnson

Contents

Viral Haemorrhagic Fevers have long captured the public imagination, conjuring up images of germ warfare, biohazard suits, and national security threats. In reality, they have a wide range of clinical manifestations and propensity to cause sustained outbreaks. What is true across all of them, is a general lack of understanding of disease mechanics, an overall lack of therapeutic treatment options, and the opportunity for spread from an animal or insect host (bat, ruminant, tick, or other) into the human population.

The bunyavirus responsible for Crimea-Congo Haemorrhagic Fever (CCHF), as exemplified by its name, has the greatest geographic distribution of known VHFs,

C. S. Brown (✉)
King's Sierra Leone Partnership, King's Centre for Global Health, King's Health Partners, King's College London, London, UK

Department of Infection, Royal Free London NHS Foundation Trust, London, UK

National Infection Service, Public Health England, London, UK
e-mail: colinbrown@doctors.net.uk

O. Johnson
King's Sierra Leone Partnership, King's Centre for Global Health, King's Health Partners, King's College London, London, UK

© Springer International Publishing AG, part of Springer Nature 2018
M. Lado (ed.), *Ebola Virus Disease*, https://doi.org/10.1007/978-3-319-94854-6_1

extending across the African continent, through Eastern Europe, the Middle East, and across to the south-east Asian subcontinent to the western edge of China. This widespread ecological niche is due to the range of the principal *Hyalomma* tick vector. Other VHFs, such as the arenavirus that causes Lassa Fever, are more limited in area, confined in this case to a few countries in West Africa where it remains endemic in rodents. The flavivirus responsible for Omsk hemorrhagic fever (OHF), is limited to muskrat population in Western Siberia. New world arenaviruses cause a variety of geographically distinct disease manifestations, such as the Machupo virus causing Bolivian haemorrhagic fever (BHF), transmitted through infected mice. The filoviruses that cause Ebola Virus Disease (EVD) and Marburg, are thought to have reservoirs in a variety of bat populations found across Africa, though most prominent in Central, East, and West Africa.

Some viruses, including those endemic in rodents and ruminant s such as those responsible for Lassa Fever and CCHF, are thought to cause a wide range of illness, from asymptomatic through a mild cold-like illness to severe disease manifestations with haemorrhagic complications and death. Others, such as types of Ebola virus including the most common Zaire strain (EBOV), are thought to near universally cause a severe illness with a very high case fatality rate if patients are left without significant, targeted supportive care.

The specific animal reservoir of Ebola virus remains unknown, though increasing evidence suggests that fruit (and likely some insect-eating) bat populations may be responsible. Bats have been demonstrated to become infected with and subsequently clear filovirus infection. Mammals such as non-human primates can be exposed an infected in identical manner to human populations, and also pose a route of introduction into a local community. The increasing nature of human ingress in natural animal habitat, coupled with food security concerns, and ease of transport both within, and between, countries and continents, means it is likely that we shall see increasing numbers of outbreaks.

EVD was first identified jointly in Zaire (now the Democratic Republic of Congo, DRC) and South Sudan in separate outbreaks with different viral subspecies in 1976.[1] Outside of a very small number of sporadic cases in the late 1970s, it was not seen, or most likely recognised, outside of a laboratory environment for the subsequent 15 years, re-emerging in Gabon in 1994. The next 20 years saw increasing numbers of outbreaks, with different species of Ebola virus (Taï Forest, Bundibugyo, and the non-human primate Reston species). The case fatality rates varied from approximately 50% to up to 90%—the largest ever seen before the West African Outbreak of 2013–2016 was in Uganda, in 2000–2001, numbering 425 cases.

The scale and severity of the 2013–2016 West African was unprecedented—by its close over 28,600 people had been affected, and over 11,300 people had died. The outbreak lasted two and a half years, involved a significant multi-agency and

---

[1] CDC. Outbreaks Chronology: Ebola Virus Diseases. Atlanta: CDC, 2017. Available at: https://www.cdc.gov/vhf/ebola/outbreaks/history/chronology.html

international relief effort at a cost of billions of US dollars (the US spent $2.4 billion alone in its response) to the global community,[2] and a devastating effect on the local economies of Guinea, Liberia and Sierra Leone.

1.1 Ebola

Ebola Virus is a single-stranded negative sense RNA virus, simple in its genetics, with only seven proteins encoded for. It produces its severe disease manifestation through a mixture of dysregulation of host immune function, causing endothelial damage, and direct viral effects on tropic tissues including the kidney, liver, brain, and pancreas. Ebola Virus Disease has an incubation period of between two-21 days, with the majority of infections occurring within the first 10 days. The initial symptoms are vague, such as fatigue, anorexia, fever, and general malaise, and mimicking a large number of infectious aetiologies including malaria. People subsequently develop gastrointestinal symptoms including abdominal pain, diarrhoea, vomiting and hiccups. Many progress to neurological involvement and some will develop abnormal bleeding and shock, followed by death. After entering the human population, it is transmitted through direct contact with infected bodily fluids (blood, faeces, vomit, semen) which enters the next person's body through contact with mucous membranes or breaks in skin. Management of the individual largely relies on symptomatic relief and correction of electrolyte imbalance, clotting abnormalities, and fluid resuscitation. Control within the community requires contact tracing of people in close contact with a case, with active symptom monitoring or possible quarantine; adequate diagnostics and isolation facilities; a trained public health and clinical workforce; community engagement and social mobilisation; and burial teams for safe burial of corpses. Following the 2014–2016, there remain no licensed treatment options, though there are promising signals for a few experimental treatments. Several vaccine candidates have been trialled, including one that demonstrated considerable effect in protecting contacts of EVD cases.

1.2 The Architecture of an EVD Outbreak Response

The clinical diagnosis and treatment of patients with EVD forms just one component of an overall outbreak response. This should be led by an Emergency Operations Centre (EOC), which national, donor, UN and non-governmental organisations should all work through. An Incident Manager should lead the EOC and provide oversight for the response. Several clusters, or pillars, should operate under the EOC

---

[2]CDC. Cost of the Ebola Epidemic: Ebola Hemorrhagic Fever. Atlanta: CDC, 2016. Available at: https://www.cdc.gov/vhf/ebola/outbreaks/2014-west-africa/cost-of-ebola.html

and these should include a configuration of: social mobilisation; surveillance; case management; laboratory services; burials; and coordination.[3]

The bedrock of an Ebola outbreak response is social mobilisation and community engagement. The public need to be informed about the virus and its transmission so that they can modify any behaviours that might put them or others at risk of infection. They also need to trust and the response, so that they agree to collaborate with it and are forthcoming with information about possible new cases. Public health messaging will be a component of this, such as radio adverts, text messaging or house-to-house visits. More critically, the response must engage with existing community structures, which may include traditional or religious leaders.

The surveillance team track the epidemiology of the outbreak and detect new cases. This may be through community-based surveillance systems, with community health workers or district health teams alert the EOC about unexplained or suspicious deaths. The surveillance team will investigate any reported cases and ensure that samples have been sent for laboratory confirmation where required. They will ensure that any close contacts of an EVD patients, who may have been exposed to the virus, are traced and monitored for the duration of their possible incubation period.

The case management team ensures the isolation and safe treatment of suspected and confirmed EVD patients. All health facilities should have robust systems in place to screen patients for EVD symptoms and to isolate in a designated room or facility. Ensuring the highest standards of infection control is an essential component of protecting health workers and preventing nosocomial transmission. One or more Ebola Treatment Unit (ETU) may need to be set up to provide separate and specialist facilities where suspected or confirmed cases can receive treatment.

The laboratory cluster ensures that diagnostic services for EVD are widely available and that testing occurs swiftly. They ensure that systems are in place for the safe packaging and transportation of samples, reliable testing, and the timely and appropriate communication of results. With the increasing availability of hospital-based or point-of-care diagnostic tests, there are now considerably more options for how to achieve this.

The burials team ensure that a safe and dignified burial is conducted for EVD patients, since unsafe burials are considered a significant route of transmission. Burials should respect local cultural practices as far as possible, whilst ensuring that they mitigate any risk of transmission from corpses. Depending on the nature of the outbreak, specific criteria must be developed for whether it is only the bodies of confirmed patients that should be included in this policy, or also suspected cases or all deaths.

Finally the coordination group should facilitate communication and collaboration between the other clusters and facilitate cross-cutting issues such as the funding, logistics and supply chain for the response.

---

[3]https://www.cdc.gov/Mmwr/preview/mmwrhtml/mm6339a5.htm

1.3   Critical Issues to a Successful EVD Response: Lessons from the West Africa EVD Outbreak in 2014–2015

The West Africa EVD Outbreak in 2014–2015 highlighted a set of critical and cross-cutting issues that are fundamental to a successful Ebola response.

Community Engagement The response must be something that is done with communities, rather than done to them. Too often, social mobilisation was siloed and deprioritised when it should have been front and centre and considered everybody's business. A significant challenge was the history of distrust of government and foreigners held by many communities. Another was that many people did not see the disease through a biomedical lens but through a spiritual one, and much of the public health messaging did not connect with traditional health believes. The response needs to understand these factors and address them.

Empathy What the response asked of communities and patients was often inhumane or unrealistic. This included, for example, telling parents not to touch their child who was sick and dying at home. For every policy or operational decision, responders should ask themselves the question: how would I feel, and what would I do, if this were me or my child?

Politics in the Response An EVD outbreak has profound implications for a country's economy, as well as for the reputation of Ministers and senior health officials. As a result, politics inevitably permeates decision-making. A particular challenge can be attempts to hide or cover-up the data about the outbreak, which can hugely undermine efforts by responders to contain it. It is essential that the EOC maintains transparency and is given the political space to make decisions that are based on the evidence and the agreed policy.

Health System Capacity The pre-existing capacity of the health system in West Africa was limited, which provided a weak foundation for the response. Many health facilities were overcrowded and lacked key infrastructure or supplies for infection control, such as running water or soap on the wards. Shortages of staff were often severe, and there was a particular lack of clinical supervisors such as consultant physicians or senior nurses, to ensure standards were maintained. There were not robust clinical systems in place, such as cohorting of high-risk patients or screening.

Staff Motivation and Accountability At the centre of this was the low level of motivation and accountability of the health workforce. In this context it was essential to ensure stronger morale and oversight amongst teams of health workers. This included setting out defining key roles for staff, ensuring that they were incentivised to work through hazard pay or risk allowance, introducing performance-based bonuses, and publicly acknowledging and celebrating success. Staff must also be assured that they will receive the best medical treatment if they become sick; the successful trial of a vaccine will likely be a game-changer for reducing health worker infections and increasing their willingness to treat EVD patients.

Command and Coordination An Ebola response involves multiple moving parts, including within case management. This includes monitoring bed capacity at different units, receiving suspected cases from the community, referring EVD-positive patients to an ETU, transporting blood samples and disseminating lab results, distributing essential supplies and liaising with burial teams. Coordinating this can place a huge burden of senior clinical staff and create chaos if multiple facilities are competing for the same support services. In this situation, it is advisable to set up a dedicated Command Centre to manage this coordination and logistics.

Quarantine and Restraint A hugely controversial issue in the response was the decision to impose district and household quarantine, which had devastating impacts on the communities but was widely considered ineffective. The introduction of voluntary quarantine facilities was one promising alternative, where high-risk contacts of EVD patients could choose to isolate themselves in a safe and comfortable environment and where they could receive regular monitoring. For health facilities, a major dilemma was whether to lock patients inside isolation facilities and physically prevent them from self-discharging. Emphasis was placed in gaining consent from patients, providing quality health care and ensuring they could communicate with relatives, so that they chose to stay. However the risk of EVD patients returning to their communities or exposing others to potential infection within the health facility was often considered to great, with many health facilities deciding to lock patients inside the unit.

Maintaining General Health Services During an Outbreak Whilst the emphasis of the response was on ending the EVD outbreak, the consequences it had on general health services were devastating. Many health services closed, patients were often afraid to come to health facilities and routine vaccinations were largely suspended. As a result, the excess mortality caused by the outbreak may be more from other conditions than from EVD itself. Maintaining essential health services is therefore critical, whilst recognising that some non-essential services, such as elective surgery, may need to be suspended. This requires excellent screening and infection control, maintaining the confidence of the public and prioritising general health services in the response. Innovative public health measures should be considered; for example, the mass-administration of malaria treatment in Sierra Leone was considered a successful way to reduce malaria mortality as well as the overall number of febrile patients in the community.

1.4   Future Outbreaks

Each outbreak will be contextual, routed in a particular geography, political context, point in time, and health system. There remain many unknowns, including transmission dynamics within households, best therapeutic options, and efficacy of vaccine across different subspecies and longitudinally. The overall operational response will always be reliant of the underlying health system, with significant challenges facing low resources settings. Nonetheless, we hope this book will frame the various facets of outbreak response, with a focus on clinical and public health management, that will be useful for those facing similar challenges in the future.

2

Marta Lado, Colin S. Brown, Naomi F. Walker, Daniel Youkee, Oliver Johnson, Andy Hall, Patrick Howlett, Hooi-Ling Harrison, Felicity Fitzgerald, and Natalie Mounter

Contents

M. Lado (⊠)
King's Sierra Leone Partnership, Freetown, Sierra Leone

King's Sierra Leone Partnership, King's Centre for Global Health, King's College London and King's Health Partners, Freetown, Sierra Leone
e-mail: mlado@pih.org

C. S. Brown
King's Sierra Leone Partnership, King's Centre for Global Health, King's Health Partners, King's College London, London, UK

Department of Infection, Royal Free London NHS Foundation Trust, London, UK

National Infection Service, Public Health England, London, UK

N. F. Walker
Department of Clinical Research, London School of Hygiene and Tropical Medicine, London, UK

Hospital for Tropical Diseases, University College London Hospital, London, UK

D. Youkee · O. Johnson · A. Hall
King's Sierra Leone Partnership, King's Centre for Global Health, King's Health Partners, King's College London, London, UK

P. Howlett
King's Sierra Leone Partnership, King's Centre for Global Health, King's College London and King's Health Partners, Freetown, Sierra Leone

H.-L. Harrison
King's Sierra Leone Partnership, King's Centre for Global Health, Weston Education Centre, London, UK

F. Fitzgerald
Infection, Immunity, Inflammation and Physiological Medicine, UCL Great Ormond Street Institute of Child Health, London, UK

N. Mounter
Solent NHS Trust, Southampton, UK

© Springer International Publishing AG, part of Springer Nature 2018
M. Lado (ed.), *Ebola Virus Disease*, https://doi.org/10.1007/978-3-319-94854-6_2

The Ebola virus and the Marburg virus together form the family of Filoviridae. The Filoviruses are thread-like RNA viruses that cause fever and haemorrhagic complications. The Filoviruses cause severe disease in humans and non-human primates (gorillas, chimpanzees and monkeys) with an extremely high case fatality rate in humans ranging from 25 to 90% depending on the subtype and the availability of medical care.

The Ebola virus is relatively fragile and vulnerable to

- Chlorine
- Alcohol and formaldehyde
- Soap
- Heat

2.1   Modes of Transmission

- *Contact with the natural reservoir or infected animals*: Humans and non-human primates can be infected after being in contact with (e.g. having touched or eaten) the unknown natural host or an infected animal. This is an uncommon way of transmission, but has to occur at least once to initiate an outbreak.
- *Direct contact with infected body fluids of an infected patient*: Contact with blood, urine, excreta, vomit, saliva, sweat, breast milk, organs, secretions and sperm (the virus can be found in semen up to 9 months after clinical recovery) can lead to infection and this is the major mode of transmission in most outbreaks.

  Routes of infection are:
  - Oral
  - Conjunctivae
  - Mucous-membrane exposure: nose and mouth
  - Sexual unprotected intercourse
  - A break in the skin
  - A penetrating object infected with body fluids of a patient, e.g. needles or razor blades.
- *Contact with infected corpses (human or animal)*: Bodies of deceased patients or animals that died of Ebola infection are highly contagious because of the high levels of virus in the corpses. Often traditional burial rituals consist of washing and touching the body to prepare the body to be returned to the ancestors. People touching and washing the corpse are at high risk to contract the disease and this is a well-documented, major mode of transmission.

- *Nosocomial transmission*: Needles, syringes and material contaminated with infected fluids, can cause infections in health staff and patients. When medical items are re-used without adequate sterilization on patients attending a health facility, numerous people and health staff can get infected. If no hand washing takes place in between consulting patients, infections can spread between health staff, and from health staff to other patients. The importance of this mode of transmission has shown to vary from outbreak to outbreak.

There is no evidence so far of airborne or aerosol transmission of Ebola Virus

2.2 Incubation Period and Communicability

The incubation period for Ebola is 2–21 days. The window period between exposure and development of symptoms is thought to be a minimum of 48 h. During the incubation period the patient is infected with the virus, but is asymptomatic and is not contagious.

During the first days of symptoms the levels of the virus increases and therefore its communicability increases rapidly. If the patient doesn't manage to establish a proper immune response, then the level of the virus continues to increase until death occurs. The corpse of a patient who died of Ebola is therefore highly contagious. If the immune response is sufficient, then the level of virus decreases gradually until recovery.

2.3 Pathophysiology

Ebola replicates in various human cells. Target cells for the virus are mononuclear cells, hepatocytes and vascular endothelia cells. Mononuclear and dendritic cells (that are involved in the immune response) are the first to be infected, leading to an immune suppression. As the disease progresses parenchyma cells, like hepatocytes and adrenal cortical cells, are infected, thereby affecting the function of liver and kidneys.

Massive release of inflammatory mediators causes an increase in vascular permeability leading to shock; and disseminated intravascular coagulation, leading to coagulopathy and bleeding. The virus can finally affect almost all organs leading to multiple organ failure (MOF) and cause widespread cell death.

The immunological response in the beginning of the infection will decide how fast the virus will multiply and if the evolution will be catastrophic or towards cure. The faster the antibodies immunoglobulin M (IgM) and Ig G appear, the more chance the patient has to survive. Death or recovery normally takes place between 10 and 14 days after onset of disease.

2.4   Clinical Features and Clinical Management

Symptoms start generally and are similar to common tropical diseases like malaria, shigellosis or typhoid. A clinical diagnosis is therefore difficult. Symptoms develop progressively and Ebola can kill rapidly.

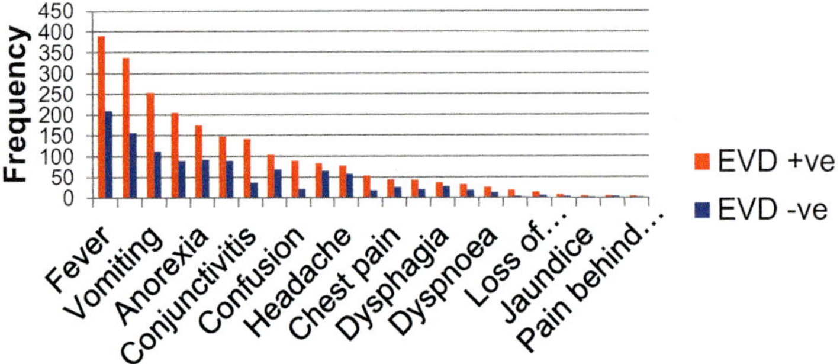

Lado, M. et al Clinical features of patients isolated for suspected Ebola virus disease at Connaught Hospital, Freetown, Sierra Leone: a retrospective cohort study Lancet Infect Dis 2015; 15: 1024–33

High probability of a confirmed EVD diagnosis

- Confusion
- Conjuntivitis
- Intense fatigue
- Hiccups
- Diarrhea

A combination of three or more symptoms
Increased the odds of EVD by 3.19% (95% CI 2.29–4.44) **Sensitivity** of **57.8%** (95% CI 52.1–62.4).

Twenty-eight percent of non cases presented with three or more, **Specificity** of **70.8%** (95% CI 64.7–76.4).

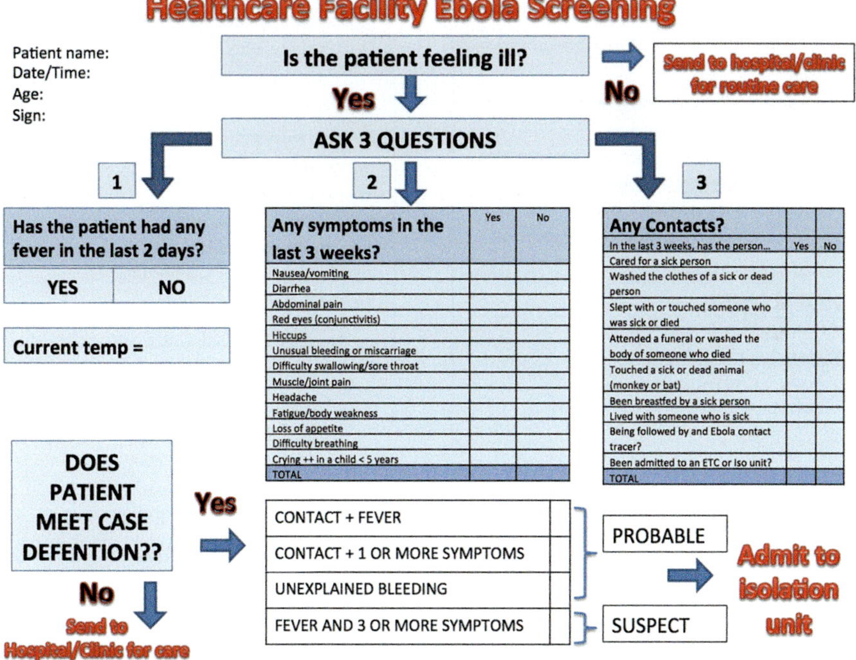

Screening tool for Hospitals during an EVD outbreak

2.4.1 Phases of Ebola Viral Disease and Clinical Presentation

Day 0-5

Pseudomalarial Syndrome

Mild fever (37.5-38.5ºC)

Body and joint pain (ocasionally back pain),

progressive weakness, loss of apetite, sore throat, headache.

Day 10-14

Hypovolemic Syndrome-Dehydration

Oligoanuria, dry mucosas, hypoglycemia,

tachypnea, tachycardia, disminished consciousness or coma.

Neurological involvement-Encephalitis

Fever, confusion and disorientation, agitation

(constant falling on the floor), bradipsichia (unable to keep atention),

extreme weakness (unable to stand up and walk), conjuntivitis.

DEATH IN 24-48h*

after the Neurological abnormalities

Haemorrhagic complications

Gums bleeding, melena, Haematemesis,

epistaxis, bleeding from punture and IV lines sites.

Day 5-10

Gastrointestinal Syndrome

Lower chest/epigastric pain, nausea and vomiting

hiccups, diarrhea (ocassionally with mucus),

crumps or diffuse abdominal pain

(sometimes liver tenderness).

>14 day:

Recovery: Resolution of GI symptoms and fever,

Increased appetite.

Convalescent weakness.

OR

Late complications: Multiorganic failure, tachypnea

(Kussmaul breathing), seizures, DEATH.

Electrolitic disfuntion

After the incubation period (between 7 and 21 days):

- **Early phase**: Mild fever, generalized weakness, headache, loss of appetite. Conjuntivitis is present in this stage. It feels like a pseudomalarial process and can be confused with that by the patients. Fever has not been detected in all patients, around 10–15% of affected patients did not refer any fever during the first phase which can make difficult the screening and detection of cases.
- **Middle phase**: the gastrointestinal symptoms start presenting with nausea, vomiting and/or diarrhoea, abdominal pain

  Patients can present severe dehydration secondary to Gastrointestinal losses (until even 10 L/day), therefore the hydration and IV fluids should be started as soon as possible to prevent this complications. Some patients complain about right upper abdominal pain (liver tenderness) or epigastric pain similar to Peptic Ulcer symptoms. Hiccupping can appear during this phase.
- **Hypovolemic phase**: due to the gastrointestinal losses and the low income of fluids, patients can suffer severe hypovolemia and be affected by renal prerenal failure.
- **Neurological phase**: many patients start presenting a neurological involvement with symptoms similar to Encephalitis with seizures, delirium and "wide eyes stare". Death is related in time with neurological complications and most of the patients that follow this pattern present death in the next 24–48 h.

- In a small percentage ($<20\%$) present hemorrhagic complications like external bleeding; more related to coagulopathy and/or thrombocytopenia.
- Multi-organic failure usually follows and the patient dies within the first 2–3 weeks since the burden of symptoms.

2.4.2 Clinical Management-Symptomatic Treatment

During the study of suspected cases of Ebola Viral disease, symptomatic brad spectrum medication should be given to cover also other possible diagnosis until the results for EVD is available.

2.4.2.1 Routine Medicines

All patients admitted to the isolation unit as suspected cases of EVD should be prescribed the following:

1. **Antibiotics** (intramuscular/intravenous Ceftriaxone)
   - Adult dose = 1 g once per day
   - Paediatric dose = (see below)
2. **Antimalarials** (oral ACT for 3 days or intramuscular/Intravenous Artesunate for three doses followed by 3 days of oral ACT)
   - Adult dose (oral ACT—Artesunate/Amodiaquine 100 mg/270 mg) = 2 tabs once per day for 3 days
   - Adult dose (intramuscular/Intravenous Artesunate) = 2.4 mg/kg at 0 h, 12 h, 24 h, followed by 3 days of oral ACT
   - Paediatric dose (see below)
3. **Paracetamol** (oral—tabs and syrup available)
   - Adult dose = 1 g four times per day (minimum of 4 h between doses)
   - Paediatric dose (see below)
4. **Oral rehydration salts**
   - Adult dose: 2 L over 24 h
   - Paediatric dose (see below)
5. **Ringer Lactate**
   - 1000 cc daily or every 12 h depending of the grade of dehydration and the oral tolerance.

2.4.2.2 Additional Medicines

Several other medicines are available and should be prescribed for patients to aid symptom control. Available medicines include:

1. Omeprazole (oral)
   - Indications = reflux
   - Adult dose = 20–40 mg once per day

2. Metoclopramide (oral and intramuscular/Intravenous)
   - Indications = vomiting
   - Adult dose = 10 mg (max three times per day)
   - Do not use in children
3. Diazepam (oral and intramuscular/intravenous)
   - Indications = agitation, seizures
   - Adult dose = 5–10 mg as required
   - Paediatric dose (see below)
4. Tramadol (oral and intramuscular/intravenous)
   - Indications = pain
   - Adult dose = 50–100 mg four times per day
   - Paediatric dose (see below)
5. Haloperidol (intramuscular/intravenous)
   - Indications = agitation, hiccups, vomiting
   - Adult dose (agitation) = 0.5 mg (max three times per day)
   - Adult dose (vomiting) = 1–2 mg intramuscular stat
6. Promethazine (oral and intramuscular/intravenous)
   - Indications = nausea, vomiting
   - Adult dose (oral) = 25 mg once per day. Can be increased to twice or three times per day (maximum 100 mg per day)
   - Adult dose (intramuscular) = 12.5–25 mg up to four times per day (minimum of 4 h apart)
   - Paediatric dose (see below)

2.4.2.3 IV Fluids

IV fluids may be lifesaving in patients with EVD.

Follow safe procedures in patients with:

- Confusion
- Agitation
- Bleeding from venepuncture site or other uncontrolled bleeding

Equipment needs

Insertion kit	Removal kit
Essential	Essential
• Ringer's lactate 1 l x2	• Metal tray
• IV giving set	• Gauze
• Metal tray	• Crepe bandage
• Sharps bin	• Good quality tape
• Tourniquet (can use clean glove)	• Disposable sheet
• Skin preparation wipes	
• Cannula	
• Gauze	
• Good quality tape	
• Crepe Bandage	

(continued)

• Disposable sheet for under arm • Saline flush if intending to cap line • Bungs (children only)	
Desirable • Tegaderm or steristrips	Desirable • Celox or equivalent haemostatic agent

For additional information on treatment of Ebola patients, see WHO (2016). Clinical management of patients with viral haemorrhagic fever: A pocket guide for the frontline health worker. Geneva.

Paediatric doses

Drug	3–5.9 kg	6–9.9 kg	10–14.9 kg	15–18.9 kg
Paracetamol (15 mg/kg)	70 mg	115 mg	200 mg	250 mg
Ceftriaxone (80 mg/kg)	360 mg	600 mg	1 g	2 g
ACT	1 tab BD	1 tab BD	1 tab BD	2 tabs BD
Artesunate (2.4 mg/kg) (im)*	0.8 ml	1.4 ml	2.4 ml	3 ml
Diazepam (fits) (0.3 mg/kg im)	1.4 mg	2.25 mg	3.75 mg	5 mg
Diazepam (agitation) (0.1 mg/kg im)	0.5 mg	0.75 mg	1.25 mg	1.75 mg
Promethazine (vomiting) (NOT in <2 year)	–	–	10–25 mg QDS PRN	10–25 mg QDS PRN
Tramadol (1 mg/kg)	4.5 mg	7.5 mg	12.5 mg	17 mg

Response to an EVD Outbreak

Marta Lado, Natalie Mounter, Daniel Youkee, and Andy Hall

3

Contents

M. Lado (✉)
King's Sierra Leone Partnership, Freetown, Sierra Leone

King's Sierra Leone Partnership, King's Centre for Global Health, King's College London and King's Health Partners, Freetown, Sierra Leone

N. Mounter
Solent NHS Trust, Southampton, UK

D. Youkee · A. Hall
King's Sierra Leone Partnership, King's Centre for Global Health, King's Health Partners, King's College London, London, UK

© Springer International Publishing AG, part of Springer Nature 2018

M. Lado (ed.), *Ebola Virus Disease*, https://doi.org/10.1007/978-3-319-94854-6_3

3.1   Isolation Unit-Ebola Holding Unit (EHU)

3.1.1   Introduction

During the West Africa Ebola outbreak 2014–2016, there were have been many different names used to describe Ebola Isolation units, including 'Holding Units', 'Isolation Units', 'Care Units' and 'Care Centres'. For the purposes of this book we use the phrase **Ebola Holding Unit** to mean a site which:

- Isolates patients who are screened or otherwise identified as suspected cases of Ebola Virus Disease (**EVD**)
- Provides initial clinical care for suspected EVD patients
- Undertakes laboratory testing to confirm the status of suspected EVD cases
- Refers positive EVD patients to an Ebola Treatment Centre (ETC/ETU) as soon as is possible
- Can safely store corpses of suspected or positive EVD patients pending collection by a burial team
- Is ideally co-located with other general health services

The size and extent of the Ebola outbreak in Sierra Leone and rest of West Africa 2014–2016 has required new responses to deliver care and safe isolation for Ebola patients.

This model has evolved from the recognition that general health facilities need to be able to screen patients and safely isolate suspected EVD cases in an Ebola Holding Unit in order to protect health workers and maintain health services. These units can play a critical role in the Ebola response by safely isolating patients away from the community and by initiating early medical treatment, thereby reducing transmission and mortality.

There is evident need to rapidly and massively scale up isolation bed capacity to help contain the outbreak, but too often the concepts of bed capacity and of absolute bed numbers have been conflated. For Ebola Holding Units, bed capacity should be considered as the number of new admissions that can be seen per week, which is far more important than the number of beds in a unit. If an Ebola Holding Unit can increase its throughput (i.e. reduce the time it takes to get a lab result for patients and to discharge or refer them, freeing up their bed for a new patient) then it can see more patients with the same number of physical beds and staff. Increasing the throughput can therefore be a far quicker, less resource-intensive and more effective way of increasing bed capacity than building additional Ebola Holding Units. Good monitoring of data and constant troubleshooting are critical for this to be achieved.

Existing health facilities can safely and effectively manage Ebola Holding Units provided support is given for training, construction and supervision. Ebola Holding Units have many strengths: pre-existing human resources and staffing structure, local leadership, functional supply chains, and engagement from the local community. This approach is not only quick and inexpensive but also sustainable; focusing on training local staff and improving healthcare infrastructure in the local community may lead to longer-term gains for basic healthcare delivery and preparedness for future outbreaks.

Key processes for setting up and running an Ebola Holding Unit include:

- Construction
- Training
- Staffing
- Facility maintenance
- Payment of risk allowance
- Provision of medical supplies and PPE
- Provision of non-medical supplies (sheets, generator fuel etc.)
- Supervision and inspection (providing technical advice and assessing safety and clinical care)

3.1.2 Layout

Ebola Holding Units normally have a 5–15 bed capacity and are ideally built within or adjacent to health facilities (such as a hospital or community health centre). They may be established within existing buildings or as tents. Key components include:

- Screening area
- PPE dressing area for staff
- Patient areas (ideally divided into 'wet' and 'dry' areas for high and low risk patients)

- Decontamination area for staff
- Incinerator and waste pit
- Body storage area
- Stores for general supplies and drugs
- Office

3.1.3 Minimum Staffing Requirements

A typical 10 bed Ebola Holding Unit would need a minimum of the following staff, divided in three shifts (with an average of 2 days off a week):

- 3 doctors or community health officers or senior nurses
- 12 care-giving staff
- 2 lab technicians
- 2 district surveillance officers
- 12 cleaning (Including security staff)

3.1.4 Patient Flow

In urban areas the target is for patients to only be admitted to the Ebola Holding Unit for 24 hours before being discharged or referred (Fig. 3.1):

Details of the flow of care throughout the whole patient journey, detailing where responsibility for each stage lies (Fig. 3.2).

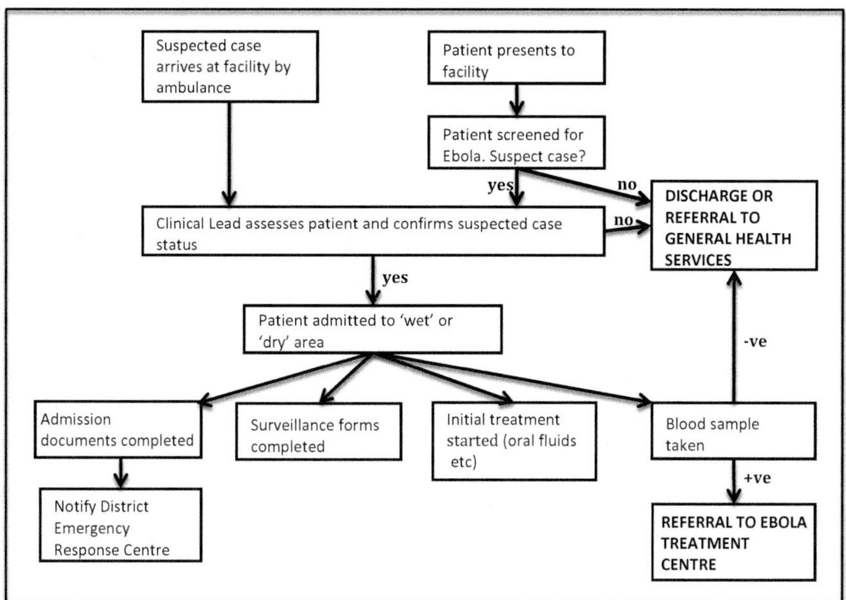

Fig. 3.1 Simplified patient flow through a unit

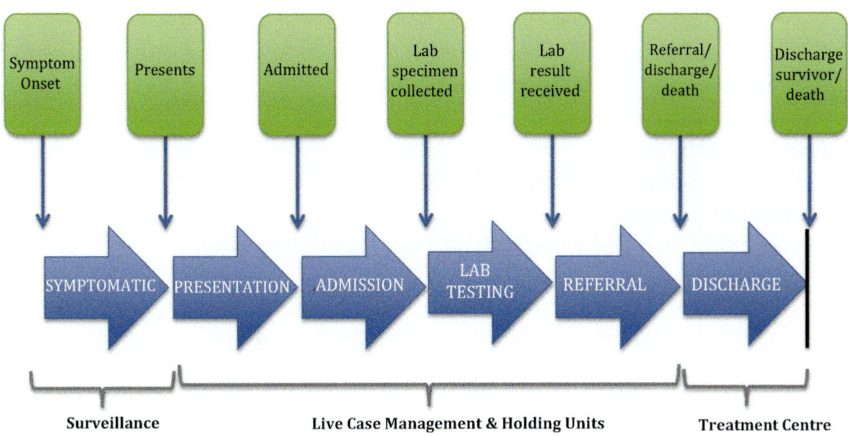

Fig. 3.2 Overall patient journey

3.1.5 Ebola Holding Units Within a Wider Framework

Ebola Holding Units do not function in isolation but require direct contact and support from others:

- **District Emergency Response Centre**: to arrange for ambulances to bring suspected EVD cases from the local community, arrange lab sample collection and results dissemination, identify treatment beds for positive EVD cases to be referred to and organise ambulances to transport them there
- **Surveillance teams**: to identify new suspected EVD cases in the community and complete surveillance forms for suspected EVD cases that present directly to the facility
- **Burial teams**: to safely collect and remove corpses
- **Central Medical Stores**: to provide regular supplies of medicines and personal protective equipment
- **General health services**: to accept non-EVD patients who require further medical care
- **Case tracking**: using the correct paperwork and case tracking mechanisms such as unique IDs is essential to track patient flow, troubleshoot issues, track individual outcomes, and provide feedback to community.

3.1.6 Site Selection

Initial site selection requirements:

- Open space for Ebola Holding Unit structure or pre-existing buildings which can be adapted

- Easy demarcation between Ebola and non-Ebola healthcare services
- Perimeter wall (ideal but not essential)
- Area for incinerator or waste disposal
- Area for screening
- Adequate water supply
- Grid energy service or ability to install generators
- Community acceptance
- Local clinical staff and availability of non-clinical workforce
- Road access

3.1.7 Staffing Requirements and Structure

The manning of an Ebola Holding Unit will differ based on its size and should take into account the number of beds/patients being managed. We suggest that staff are recruited and sourced by the Unit Coordinator, ideally from within the health facility, and then assessed for suitability for roles by the Supervising Partner, which is faster and avoids tensions within the workforce.

The following structure is based on a 4-shift system. This can be scaled up or down based on unit size.

Shift Patterns

A	08:00–14:00
B	14:00–20:00
C	20:00–08:00
D	08:00–18:00

Manning

(a) **Unit Coordinator (Shift D)**
    Figure of authority within the existing clinic who is willing to take over running of the unit full-time. Previous experience running health care facilities is highly desirable.

(b) **Unit Clinical Lead (Shift D)**
    Background in clinical medicine required. Must be willing to enter patent areas inside of the Ebola Holding Unit regularly. Can be the same as the Unit Coordinator.

(c) **12 care giving staff (Shifts A, B and C)**
    Preferably from a nursing background. Must be capable of simple protocols and understand basic infection control. Must be able to administer IV, IM and PO medication. This allows for a shift pattern of 3 early, 3 late, 3 night, aiming for 1–2 day off per week per staff member.

(d) **12 cleaners/hygienists (Shifts A, B and C)**
     This allows for a shift pattern of 3 early, 3 late, 3 night, aiming for 1–2 day off
     per week per staff member.
(e) **>2 Security (Shifts A, B and C)**
     The number of local security staff required will be dependent on the area and
     the unit. There must be at least one security staff member present all times.
(f) **District Surveillance Officer (DSO) (Shift D)**
     This individual is key to the accurate registration of patients and reporting to
     central command structures within the district (where they exist), and must be
     incorporated in to unit opening plans. If the existing patient registration and
     command and control structures that exist in the Western Area continue to be
     utilised across the country, good DSO coverage of new units will be essential.
(g) **2 Laboratory staff (Shift D)**
     These staff are in charge of collecting blood samples and/or swabs of
     suspected cases.

3.1.8 Set up and Construction

Ebola Holding Units will change in lay out from one location to another depending
on the infrastructure available.

3.1.9 Classification of Areas

The terminology often used is:

- **'Green Zone'** for the areas where there is no patient contact and no risk to staff of
  exposure,
- **'Red Zone'** for the areas inside the Holding Unit where there are patients and
  infectious waste.

The Red Zone encompasses all aspects of waste management including the
disposal of infectious material and the storage of bodies.

The flow of the unit must ensure that all staff are dressed in full PPE before entry
into the Red Zone, and are fully decontaminated on exit. Nothing should leave the
Red Zone without full sterilisation.

If this is not possible, all material must be destroyed. Patients who test negative
should have a shower before exit.

3.1.10 Basic Requirements

All facilities should meet the requirements outlined below as a minimum:

a. The area should be well ventilated to reduce heat and humidity and to allow the evacuation of chlorine gas.
b. Units should have a simple design to encourage the flow of staff from low to high-risk areas.
c. A minimum of one clean entrance, one dirty exit and one clean exit is suggested.
d. A channel to pass material or drugs from the Green area to the Red area to prevent manipulation of high-risk sharp utensils within the Red Zone. It should be a one way direction, no material should be moved from the Red Zone to the green one.
e. Internal doors between Red and Green Zone areas should be secured to prevent the movement of patients and possible contamination of decontamination areas or Green Zones
f. Each bed should have separate designated toilet, or individual latrine pot/bucket, to prevent transmission between patients and "Cross Infection". Because urine and faeces are highly contagious and infectious body fluids, safe disposal of this sewage is essential. Note that existing sewage systems may drain into neighbouring structures, which could risk contaminating them and infecting others in the community.
g. The Ebola Holding Unit perimeter should be secured from non-EVD patient or visitor incursion and EVD patient excursion. To protect staff, non-EVD patients and health facility visitors, the unit may need to be locked, barricaded, or otherwise secured.
h. The unit should be spacious enough to accommodate clean and dirty PPE changing stations, patient care with separation (see Figs. 3.3 and 3.4), waste collection and disposal (including liquid and dry waste, burn pits, incinerators, etc.), storage of corpses, laundry, safe water and chlorine solution preparation.
i. The patient care room within the unit should be spacious enough to separate patients by at least 1.5–2 m. Screens should be used between patients to maintain privacy, discourage patient movement, and reduce transmission between patients. Screens should be made of materials that can be easily disinfected (e.g. plastic) and preferably semi transparent to allow light, but also provide an element of privacy.
j. Suspected EVD cases should be separated from confirmed EVD cases whenever possible.
k. Efforts should also be made to separate 'wet' patients (those producing bodily fluids, like vomit and faeces) from 'dry' patients (those not producing bodily fluids), although this may not be possible if the unit is normally full.

Example Layouts

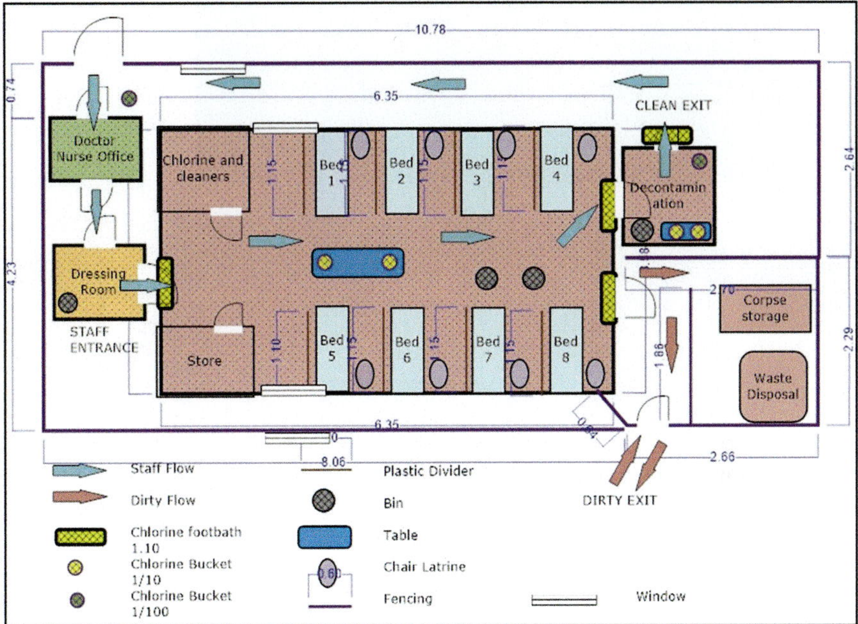

Fig. 3.3 Example of simple unit: one room or tent version. Beds 5, 6, 7, 8 can be designated for wet patients and movable barriers can be used to help delineate staff flows through the unit. Preferably if space allows there should be two separate rooms

3.1.11 Special Structures

Tents may be used for rapid deployment. Several considerations should be taken in to account:

- **Environment**: tents will be subject to direct sunlight throughout the dry season and heavy rainfall during the wet season. The internal temperature can rise to unsafe levels for both patients and staff. As a result it is advised that a sloping roof structure is built above the tent to provide shade and additional cover from rain.
- **Ventilation**: tents must have ventilation flaps, which should be opened during chlorine cleaning of the unit.
- **Doorways**: if possible tents with formal doorway structures rather than foldable cloth entrances are advised, as flaps pose a contamination risk when opening/ closing/entering/exiting.
- **Perimeter fence**: if a tent structure is used it is difficult to secure the entrance and exits, so all tent structures should have a perimeter fence built around the tent.

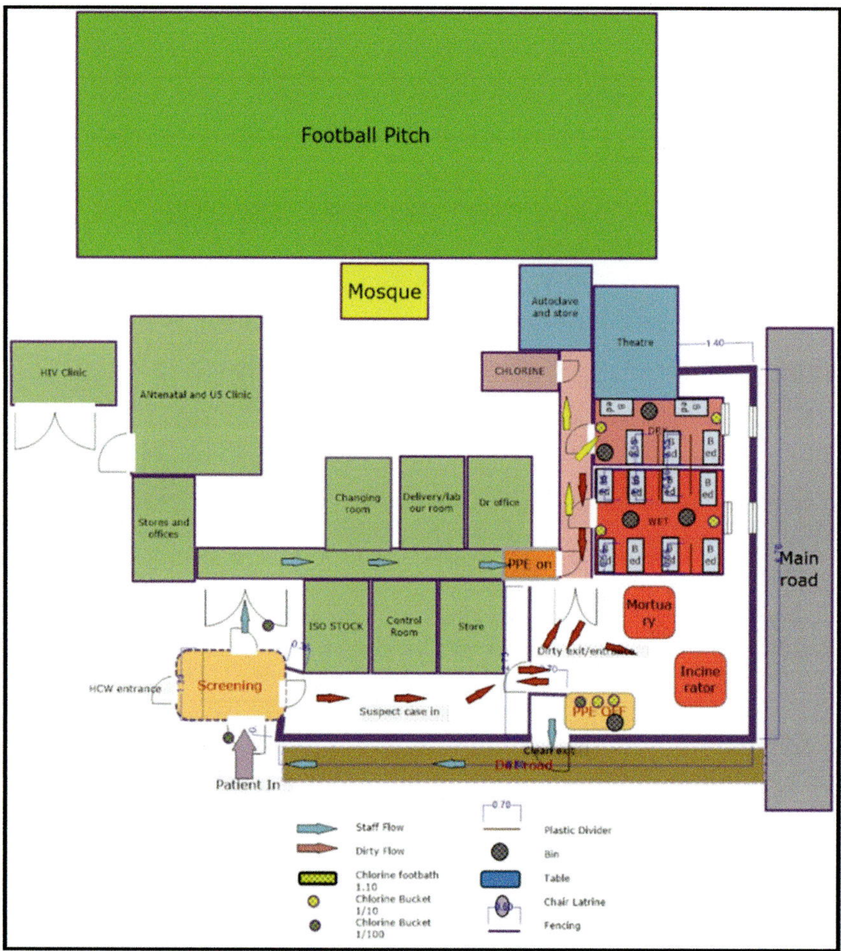

Fig. 3.4 Example of adaptation of existing building, based on the Ebola Holding Unit at Lumley Hospital, Freetown. Sierra Leone. Pre-existing rooms such as those labelled in blue should be securely blocked off

- **Foundations**: tents should be erected on solid foundations suitable for cleaning, ideally concrete with a sufficient grade to drain effluent into a soak away/French drain.

A rapid, cost effective, safe and pragmatic solution for Ebola Holding Units that cannot adapt existing structures would be a satellite-site constructed with UK Armed Forces Temporary Deployable Accommodation (TDA). These modular units can be transported and erected safely and rapidly in remote and urban areas alike. They are large enough as single modules to accommodate 6–8 patients comfortably. These should use existing Rollertrack flooring with linoleum cover available, with additions of an incinerator unit and water tank, plus 2–3 smaller tents for office, dressing and undressing. Simple units can be transported to locations and erected within 24–72 h dependent on the size of the unit. It can be helpful to erect them close to existing sites so that staffing and infrastructure can be shared.

3.1.12 Waste Disposal

- Existing incinerators are usually not sufficient to cope with the increase in waste.
- Storage of waste is not appropriate and all waste must be burnt as soon as is practicable.
- Waste disposal should be undertaken by staff fully trained in wearing and removing safely full PPE.
- The major factors to consider when disposing of waste are both quantity and type.
- Wet and dry waste will both need to be burned.

In order to effectively burn wet waste, both solid fuel (e.g. wood) and liquid accelerant such (e.g. petrol or kerosene) will be needed. The fire will need to be established before adding the wet waste and dry waste, alternating between the two to maintain an appropriate temperature to kill all infective pathogens.

Several methods can be used to burn this waste. The most common and easiest to construct is a burns pit. Simple $6 \times 6 \times 6$ foot pits are sufficient for small amounts (e.g. waste from a six bed unit). For larger volumes of waste a more sophisticated burns pit will need to be built that gives more consideration to air supply and ventilation, as well as protection from the elements during the rainy season.

Construction of flues and chimneys may be needed for larger pits. Constructed from corrugated iron and perforated to allow entry of air to the bottom of the pit, these can keep the supply of clean air sufficient to maintain heat and intensity.

In order to maintain safety a barrier will need to be constructed around the top of the pit to avoid accidental falls.

3.1.13 Mortuary

A mortuary should be set up for storage of corpses. Removal of corpses from the ward should be done as per the procedures outlined in Sect. 3.3. Rapid removal of corpses is essential to prevent cross infection and ensure dignity and safety is maintained in the clinical area. It also frees up beds for new suspected EVD cases.

A designated room or an outside roofed structure can be used as a mortuary. Corpses should be stored off the ground in body bags. A soak away should be in place if possible to allow for easier cleaning of the area.

3.1.14 Stocks and Supplies

3.1.14.1 Stock Room

During Stage 1 a suitable room needs to be identified and prepared for use. The room needs to be secure and it must be clear who is in possession of all the keys. The availability of key holders across all shifts needs to be considered and plans put in place to make sure that the unit does not run out of stock when the key holders are absent. If there is doubt about the integrity of key holders it might be necessary to change the locks and distribute keys to designated staff.

The room must also be dry, well ventilated and free of vermin. It may be necessary to clear the room of general hospital supplies before starting to stock supplies for the unit. Pallets can be useful for storing boxes on if the floor is damp. Staff should be encouraged to arrange the supplies in a logical and tidy manner, keeping supplies of like items in the same area such that staff can easily access the full range of items with ease.

Supplies should be grouped according to size (e.g. small sized gloves vs. medium sized gloves etc.). Boxes should be packed so that labels can be seen easily without having to move the boxes. For drugs, expiry dates should be observed and those items with the shortest expiry dates should be placed in at the front of the store so that they will be picked up and used first.

Stock cards are important for recording stock received and stock issued. An example stock card can be seen in Appendix 1. This helps with calculating consumption rates and knowing how much stock to order. The cards should also include a balance column so that the amount remaining can easily be seen without having to manually recount the stock. All staff that will have access to the store room must be instructed in how to record stock received and stock issued. It is good practice to do a weekly inventory check to compare the quantity listed on the stock card with the physical stock. The Supplies Supervisor can discuss any discrepancies with the staff to improve practice.

3.1.14.2 Supplies

During Stage 1 the following supplies will need to be acquired:

- **Set-up items** (e.g. buckets, footbaths, mops, brooms etc.)—see Appendix 2 for more details
- **Medical supplies** from the Central Medical Stores/District Stores—estimated quantities for a 20-bed unit are shown in Appendix 2. Orders for the Central Medical Stores/District Stores should be put in writing and have to be signed by the most senior member of staff of the health facility. It is recommended that there is a discussion with the existing health care staff about the usual processes for

signing orders and their submission to the Central Medical Stores/District Stores (e.g. whether verbal orders would be accepted, lead in times for delivery etc.) and a discussion with the stores about acquisitions, as there are different processes in place for Ebola Holding Units. It is helpful to include desired quantities for each item ordered.

- **Non-medical consumables**—see Appendix 2 for a guide to useful items and quantities.

3.1.14.3 Storage Within the Unit

In addition to the store room for bulk supplies, it is important to have the following storage spaces within the unit in order to reduce the interaction between staff inside and outside the unit and reduce the risk of supplies becoming contaminated inside the unit:

- Shelving in the dressing room for storage of PPE
- Storage cupboard or room within the Green Zone (similar to a Pharmacy) for storage of drugs, drinking water and other consumable items. This could also be an area where injectable drugs are prepared if there is equipment to do it. With an easy communication with the Red Zone in case those medications are urgently needed to be prepared and supplied to the staff inside.

These areas should be stocked with enough supplies to last for a 24-h period. A stock list should be devised to specify the minimum stock level needed for the unit to run at full capacity for 24 h. At the end of the night shift, or the start of the early shift, one nurse should check the quantity of stock physically present against the stock levels on the stock lists, and identify the amount of stock that needs to be ordered to ensure that this room is topped up to the minimum stock level.

The stock to be ordered should be communicated to staff in charge of the stocking, who note down the order on an order form and arrange for the required items to be obtained from the bulk store room. A similar stock list system can be used for the paperwork kept in the office area to ensure that these documents are available at all times.

Example stock lists and order forms can be found in Appendix 1 and 2.

3.1.15 Case Study: Lumley Community Hospital, Freetown. Sierra Leone

Location

Lumley Community Hospital is situated in the heart of Lumley Town, adjacent to the main road and football pitch. It was a functioning community hospital until the EVD outbreak reached Freetown. A serious reduction in both healthcare services offered and staff attendance was catalysed by the infection and subsequent death of a prominent doctor at the facility.

The site had previously hosted a 2-bed isolation unit. This unit was not currently functioning at the time of KSLP's assessment and was poorly designed.

History
Previous services and facility capabilities:

- 16 bed inpatient wards, separated in to male and female
- Antenatal clinic
- Delivery room
- Operating theatre for minor surgical operations
- Under 5 years old children clinic
- HIV clinic
- Laboratory for basic microscopy and blood tests
- General practice/OPD clinic

Staffing
Over 100 clinical and non-clinical staff.

Services
At the time of the initial site visit, the only healthcare services being offered at the site were the antenatal clinic, the Under 5s clinic and the HIV clinic. All other services were closed and staff was not attending work. One wing of the hospital, where the office of the doctor who died was located, had been effectively closed off and the facilities in this part of the hospital were not functioning.

3.1.15.1 Implementing the Three Stage Process

Approval
KSLP were formally requested to support the site by the District Emergency Response Centre. KSLP then visited the site on 24th October 2014 to work with the Medical Superintendent to gain consent and jointly formulate a plan to develop an Ebola Holding Unit. The installation of the Ebola Holding Unit and patient screening procedure would aim to maintain the functioning of the services still operating and encourage the re-opening of the general practice/OPD clinic.

Stage 1: 24.10.14–30.10.14

Planning
The initial Stage 1 visit was performed on 25th October 2014 and detailed plans drawn up for the construction partner (GOAL). GOAL commenced work on 26th October 2014 and the unit construction was completed on 30th October 2014.

Awareness of the local context and the current functioning of the facility lead to the proposed construction shown below. The office of the former doctor was reinvigorated by turning it into the unit control room and the disused wards were deemed easily adaptable to a high quality Ebola Holding Unit.

Knowledge acquired of existing staffing levels and capabilities lead to the decision that the new unit could host 12 adult beds with a possible 2 paediatric cots.

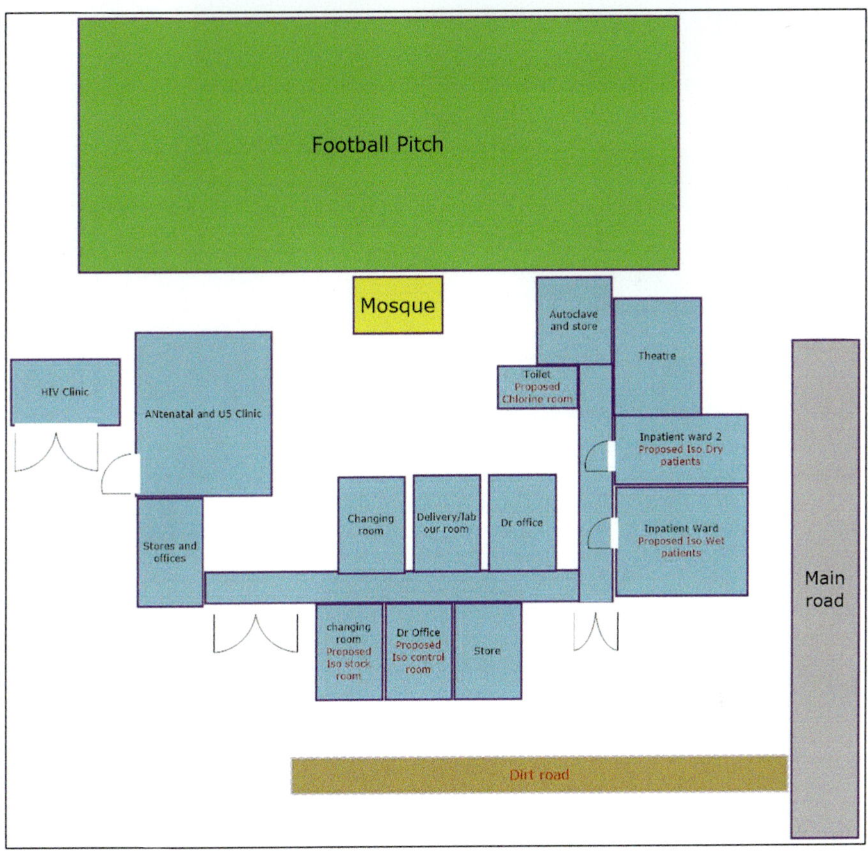

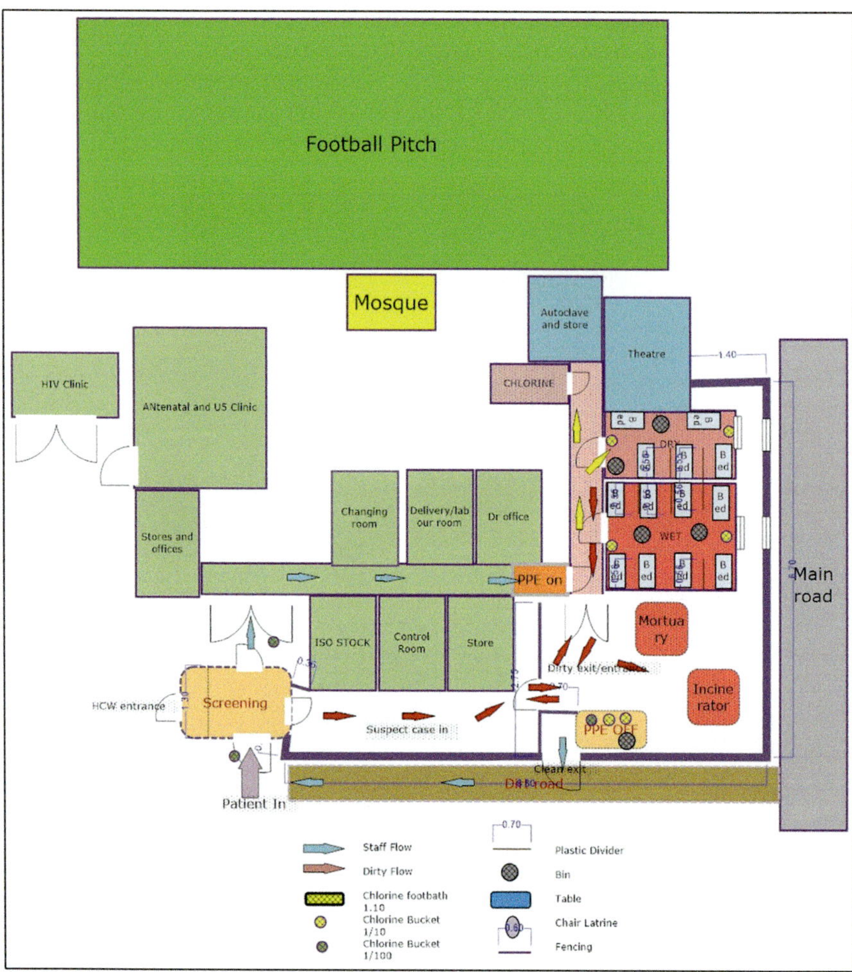

Construction

- **A deep pit incinerator** was dug in the outside space, fitted with flues for proper air circulation. A barrier was built around to prevent staff from falling and sandbags used to stop water flow in to the pit. A roof was built to cover the pit and prevent the burning of surrounding trees and structures.

- **External fencing** was placed along the road, with two meshed windows installed to allow communication between patients and relatives. The external fencing was vitally important in this setting, due to a lack of perimeter walls and the close proximity of the hospital to the surrounding community. The fencing was continued around the outside of the site to the screening area. A double door was installed to allow entrance of ambulance and burial teams.

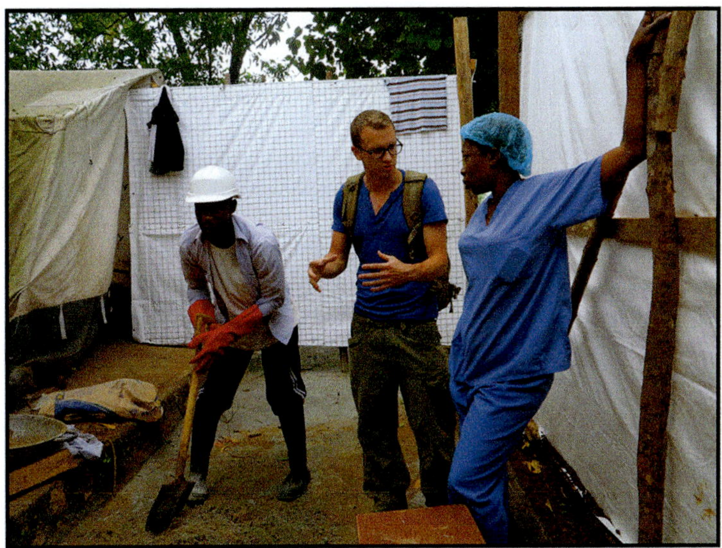

- A **screening area** was constructed with four entrances: patient entrance, HCW entrance, non-suspect case exit and suspected EVD case exit.

- **Soak aways** were constructed for proper drainage of outside areas.
- A simple roofed four-walled **mortuary** structure was built.
- The tent used for the previous, smaller unit was adapted for decontamination purposes.
- The **two existing wards were redesigned** as shown in the diagrams above.
  - Plastic sheeting from floor to ceiling was placed between each bed. Mattresses were covered in waterproof mattress covers.
  - Curtains were removed and windows cleaned to allow adequate light in to facility.
  - Surplus areas such as the operating theatre and autoclave rooms were locked.
  - Internal doors were built to turn corridor space in to dressing room and secure entrance in to Red Zone.
  - Bedside tables were placed next to beds for easy storage of patient food and ORS.
  - Latrine chairs were installed next to each patient's bed.

Staffing

All Red Zone staff was sent to the Ebola Holding Unit at Connaught Hospital for Phase 2 training (real life experience of working in a unit) between 25th October 2014 and 30th October 2014. A total of nine nurses, four laboratory technicians and six cleaners were trained. All clinical staff had previously received Phase 1 training. The site benefited from already having three security staff plus RSLAF (Republic of Sierra Leone Armed forces) presence.

From amongst the clinical staff a senior nurse was selected as Unit Coordinator (who also assumed Unit Clinical Lead responsibilities), another senior nurse was selected as Deputy Coordinator, and a third nurse was given the responsibility of coordinating the late shift.

Supply

The Supplies Supervisor visited the site on 29th October 2014 and current stocks and supplies were assessed. A stock room was secured and cleaned, stock cupboards and lists were put in place, and an urgent order was placed to Central Medical Stores. Training on stock management was given to relevant staff.

Coming 'online'

The District Emergency Response Centre was informed about the unit's opening date on 27th October 2014. Prepaid phones were supplied to the unit, contact details and operational protocols were shared and taught to the unit staff. Two DSOs were assigned to the site.

Administration

The new office was supplied with patient whiteboard and all necessary case-related documents.

Stage 2: 31.10.14–06.10.14

Timeline

31/10/14	AM, dry run. PM First two patients admitted
01/10/14	Two further patients admitted
02/10/14	One positive EVD case transferred, one patient discharged, two further patients (walk-ins) admitted
03/10/14	Two patients admitted
04/10/14	Two deaths, no further patients admitted
05/10/14	Two patients admitted
06/10/14	Two patients discharged, two patients admitted. One patient who tested negative still remained symptomatic; on closer history and examination he was sent to Lumley HIV clinic where a positive HIV test was obtained with a low CD4 count

KSLP closely controlled the number of admissions and incremental increase in bed capacity, as demonstrated in the timeline above. See Clinical Supervision Checklist (Appendix 4) for details of the supervision that took place.

Events

There were three major events that required immediate responses within Stage 2:

1. **Community tension and stigma amongst healthcare staff**. The unit staff were discriminated against and barred from entering non-EVD areas of the healthcare facility (common room/toileting area).
   Response: A staff meeting chaired by the Medical Superintendent was arranged for the next day with KSLP attendance. Education was provided along with assurance that the decontamination process within the unit was rigorous. The need for professionalism and solidarity amongst all healthcare staff was enforced.
2. **First two patient deaths**, which, similar to previous experience at other Ebola Holding Units had a negative impact on staff. In this case, staff argued over whose responsibility the bagging and removal of the corpse to the mortuary area was, and underlying fear aggravated tensions. **Response**: KSLP staff selected one cleaner and one nurse and took the opportunity to remove the corpse with their help and demonstrate the correct implementation of the protocol. The group discussed this afterwards and decided that in future all staff would share the responsibilities of corpse removal.
3. **Construction failure**. An internal door leading to the 'dry' patient area fell off its hinges, after being forced by a patient. This allowed the possibility of patients moving freely around the unit.
   Response: An immediate solution was found by installing temporary bolts to restrict access. The Facility Supervisor then visited the site the next day to install permanent fixings.

Additional Improvements During Stage 2

1. The unit was being supplied by the NPA (national) grid with the addition of pre-existing lights. Whilst throughout Stage 1 the NPA lighting was appropriate for use, during Stage 2 the low current meant inadequate lighting was present in

one ward of the unit. A generator was purchased the next day and rigged in to the existing supply as back up for when NPA current was insufficient. A letter was also written by the Medical Superintendent, and at her request countersigned by KSLP, to be presented to the NPA to request immediate improvement of the electricity supply to the hospital.
2. Throughout Stage 2 advice and mentorship was given to all staff. Staff roles and responsibilities were adapted (increased or decreased) after assessment of performance and individual skills. A natural leader was identified amongst the cleaning staff and given overall responsibility for the cleaning duties.
3. Throughout Stage 2 regular meetings were held with community chiefs and non-EVD healthcare staff. If possible the Clinical Supervisor should be present at meetings to gain awareness of the local context and mediate possible disputes

3.2   Infection Prevention and Control

Infection Prevention and Control (IPC) is any activity that is undertaken to prevent the spread (transmission) of infection between patients and staff, patients and other patients, staff and patients, contamination of the facility environment.

Summary of infection control measures for EBOLA include:

- Isolating patients
- Restricting movement of patients, staff and materials (i.e. control flow) into, within and out of the isolation unit
- Personal Protective Equipment (PPE)
- Cleaning and disinfection
- Safe transport of patients
- Safe management of corpses
- Appropriate handling and disposal of sharps and other infectious waste
- Safe Work Practices

3.2.1   Hand Hygiene

Effective and regular hand hygiene is essential, both inside and outside of the isolation unit. Ensure nails are clipped short, and remove rings, bracelets and wristwatches. Hands should be regularly washed using the steps outlined in Fig. 3.5.

Whilst inside the isolation unit, an **additional pair of outer gloves** must be worn for patient contact, changing the gloves and performing hand hygiene immediately before and after each patient contact using the buckets of chlorine (1 in 10). In this

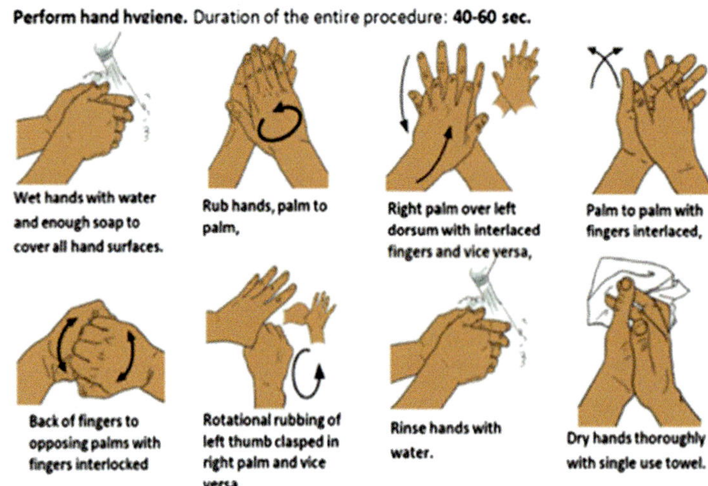

Fig. 3.5 Hand washing steps

way you will not transmit the infection from patient to patient, or patient to inanimate object, or from the object to the patient.

3.2.2 Personal Protective Equipment (PPE)

The use of PPE is to ensure the safety of frontline workers. However, this can only be achieved if:

- PPE is put on correctly
- Avoiding any readjustment after PPE is on and you are inside the Red Zone
- Ensuring PPE is removed correctly (this is the most important aspect)

  Common PPE mistakes/hazards include:

- Uncovering wrists with wide movement, especially in tall people
- Readjusting PPE (especially face shields/goggles) with contaminated gloves
- Early removal during removal procedure that increase risk of contaminating face
- For people wearing medical glasses, increased risk of glasses falling when removing the face shield (secure glasses by tying)
- Patients (especially children) touching or pulling the inside of visors

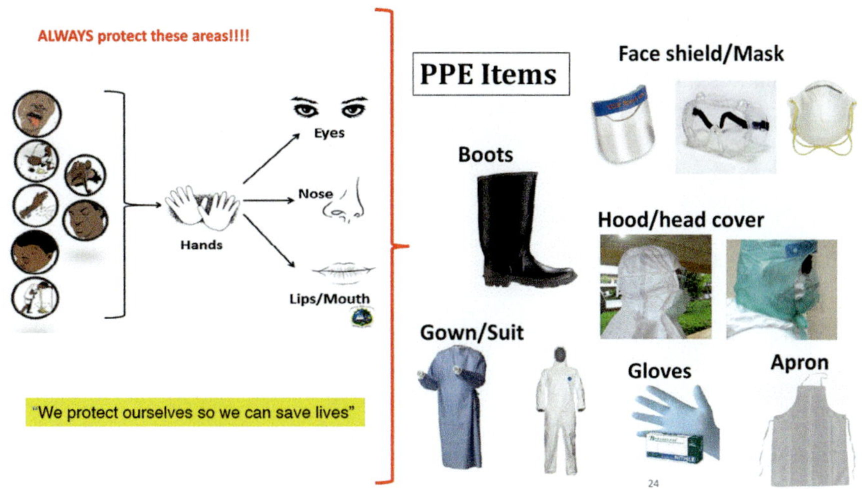

Principles of Personal protective Equipment

3.2.3 PPE on Protocol

Please follow this protocol every time you enter the RED ZONE. Additionally if entering any other area assessed as high risk, difficult transfer of suspect cases or cleaning high-risk areas.

1. Ensure any cuts/wounds are covered
2. Ensure that decontamination chlorine has been made up, especially in early morning hours.
3. Put on shoe covers and gloves
4. Enter dressing room
5. PPE SUIT WITHOUT ATTACHED SHOE COVERS
   a. Put on PPE suit over gloves, make small hole for gloved thumb in suit. Ensure PPE suit zipped up fully to neck
   b. Remove street shoes, put on foot covers, and then boots, pull PPE suit over boots.
6. PPE SUIT WITH ATTACHED SHOE COVERS
   a. Remove street shoes, put on foot covers, and then put on boots
   b. Put on PPE suit over gloves, make small hole for gloved thumb in suit. Ensure PPE suit zipped up fully to neck
7. Put on second pair of gloves over PPE suit
8. Put on apron

9. Put on face shield, ensure securely placed, with suit covering lower part of face shield, pull hood of suit up securing visor at the sides and leaving small ventilation gap at top of visor.
10. Check with buddy that PPE is correct
11. Enter the unit.

3.2.4   PPE Off Protocol

(refer to poster below, which is also displayed inside the decontamination room)

1. Ensure sprayer located next to chlorine footbath at exit.
2. Enter Decontamination room
3. Ensure space available in bin, if not use stick to push gently down, or empty bin.
4. Wash hands with soap and water in the sink to remove solid waste material on hands before decontamination
5. Wash gloves hands in 0.5% chlorine solution for 1 min (**First bucket**).
6. Remove apron by grasping at sides and holding away from you deposit in bin.
7. Wash gloved hands in 0.5% chlorine solution for 1 min (**First bucket**).
8. Remove outer gloves, deposit in bin.
9. Wash gloved hands in 0.5% chlorine solution for 30 s (**Second bucket**).
10. Looking in mirror take zip of PPE suit between thumb and forefinger of one hand and undo. Grasp hood of suit and pull off head. Remove suit, touching as little as possible.
11. Wash gloved hands in 0.5% chlorine solution for 30 s (**Second bucket**).
12. Looking in mirror remove face shield.
13. Wash gloved hands in 0.5% chlorine solution for 30 s (**Second bucket**).
14. Remove inner gloves.
15. Wash hands in 0.05% chlorine for 1 min (**third bucket**).
16. Put on new pair of gloves
17. Proceed to chlorine footbath
18. Stand in chlorine footbath
19. Spray boots with sprayer.
20. Leave isolation unit through back door
21. Remove gloves
22. Wash hands in 0.05% chlorine for 1 min (**fourth bucket**).
23. Put on new pair of gloves
24. Return to dressing room
25. Remove boots and place in boot rack
26. Put on shoes
27. Remove gloves and place in bin

28. Wash hands with soap and water
29. Put on gloves and remove shoe covers
30. Leave dressing room
31. Apply alcohol gel

3.2.4.1 Removing PPE: 20 Steps to Save Your Life

Pre-Phase 1

Step A. Wash obviously soiled gloves in soap and water

Phase 1: Bucket 1

Step 1: Wash gloved hands for 1 min (1/10)

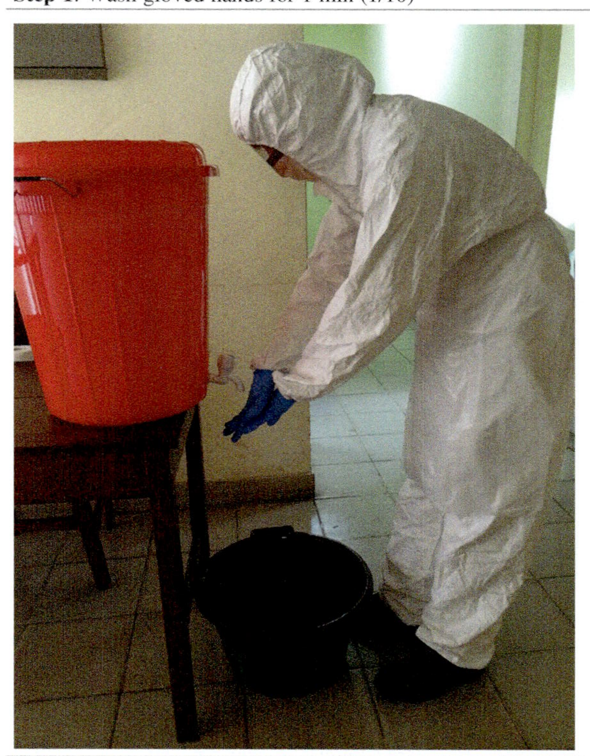

Step 2: Remove apron Dispose in bin

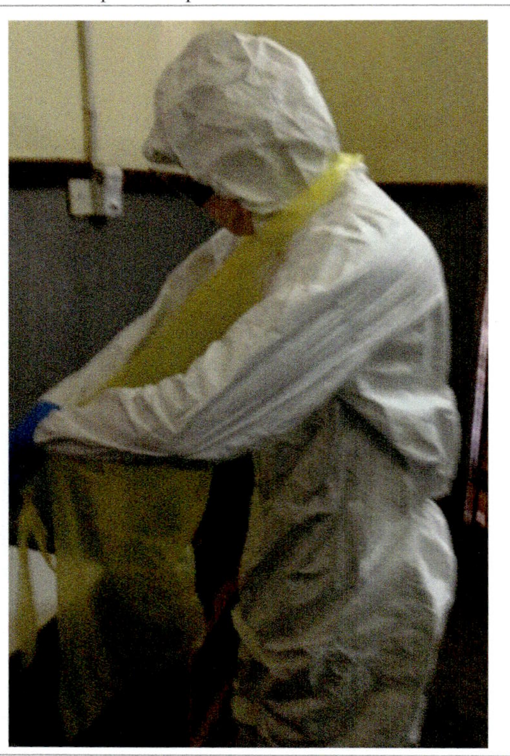

Step 4: Remove outer gloves Dispose in bin

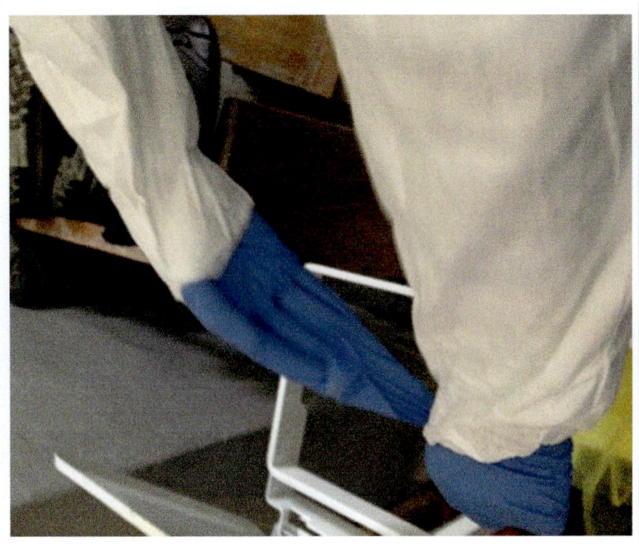

Step 3: Wash gloved hands for 1 min (1/10)

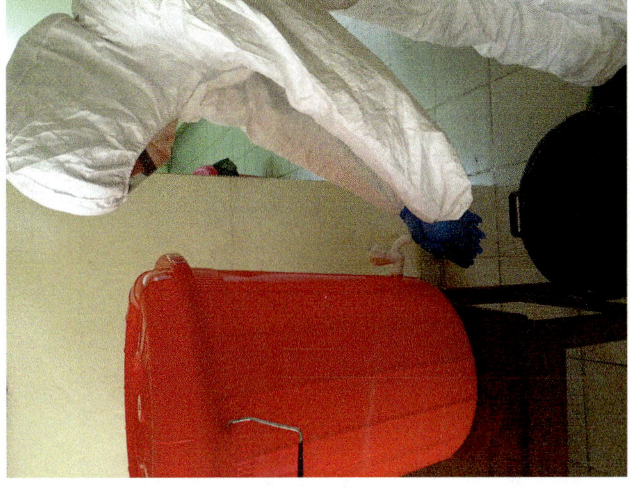

Phase 2: Bucket 2

Step 5: Wash gloved hands for 30 s (1/10)	**Step 6**: Use mirror remove suit, Roll the inside out as you remove. Dispose in bin
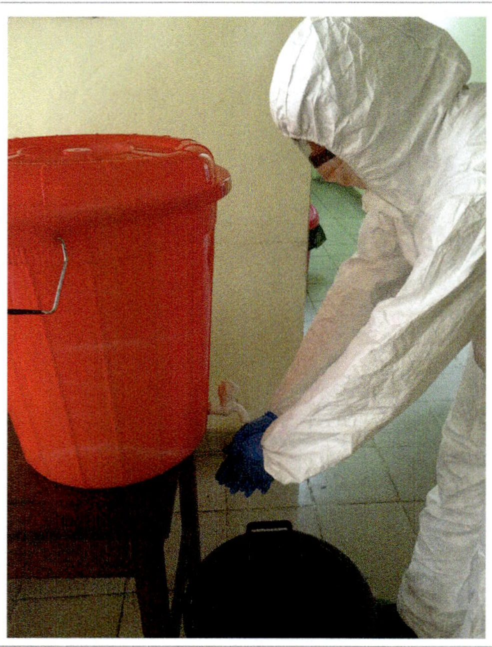	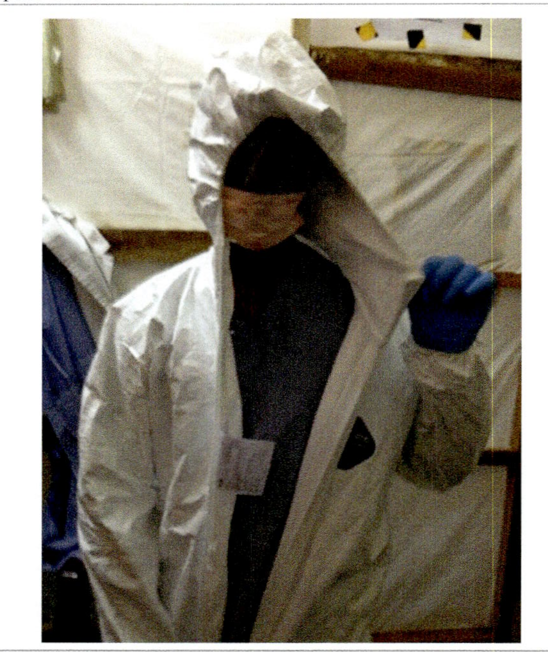

Step 7: Wash gloved hands for 30 s (1/10)	**Step 8**: Use mirror to remove facemask from the sides. Close eyes as removing. Don't touch front part. Dispose in bin
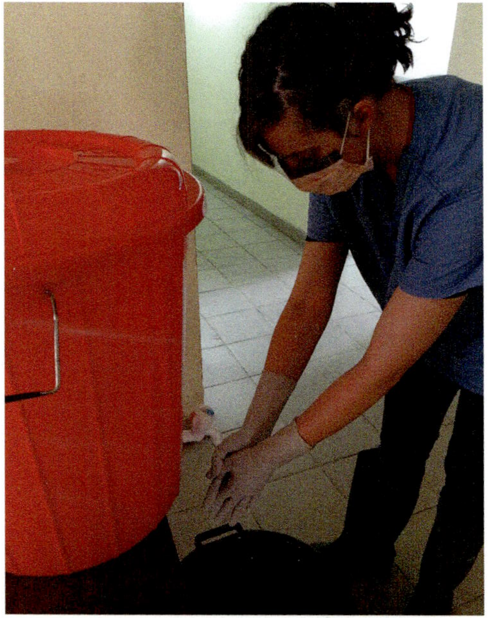	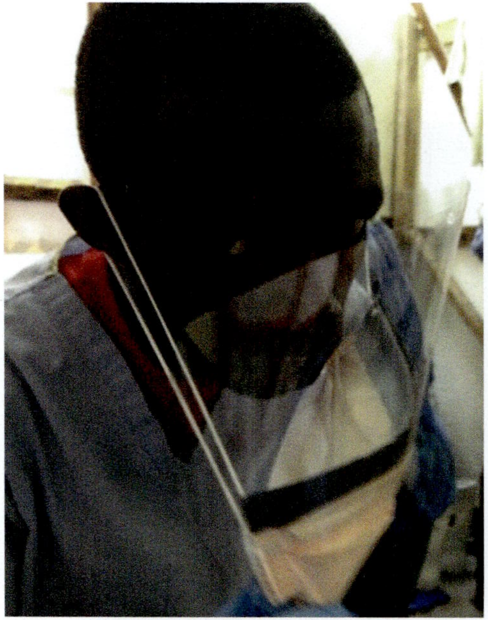

(continued)

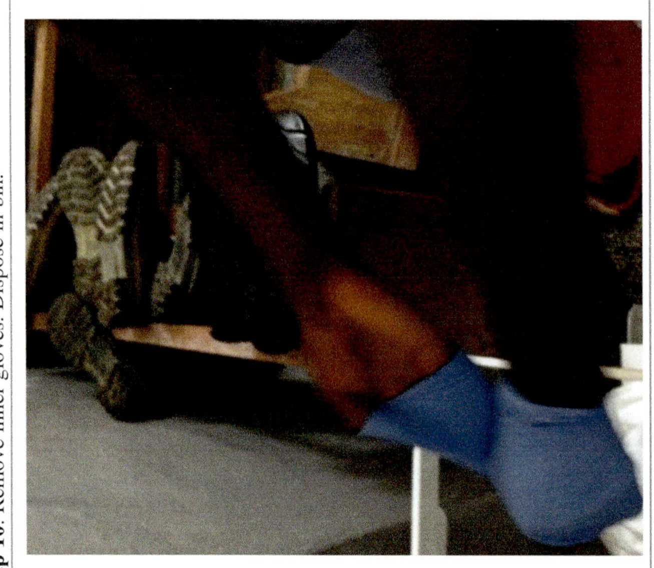

Step 10: Remove inner gloves. Dispose in bin.

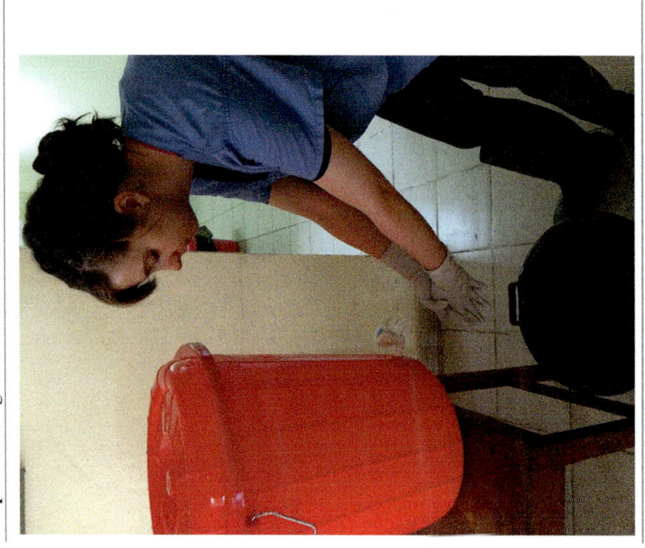

Step 9: Wash gloved hands for 30 s (1/10)

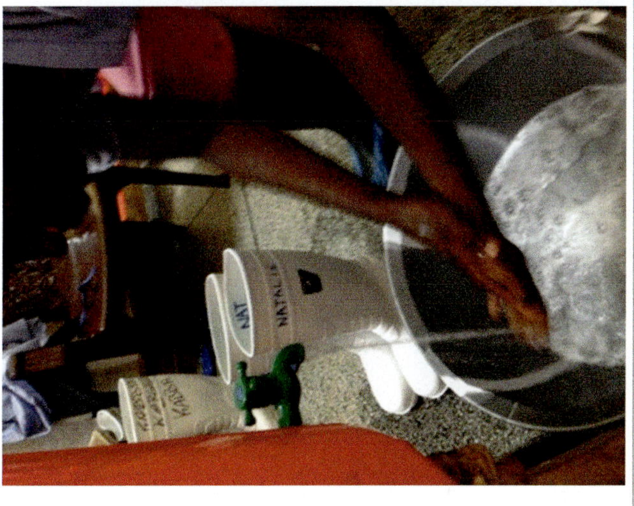

Step 11: Wash hands for 1 min. Wash up to elbows (**1/100**)

Phase 3: Boot Cleaning

Step 12: Put on gloves

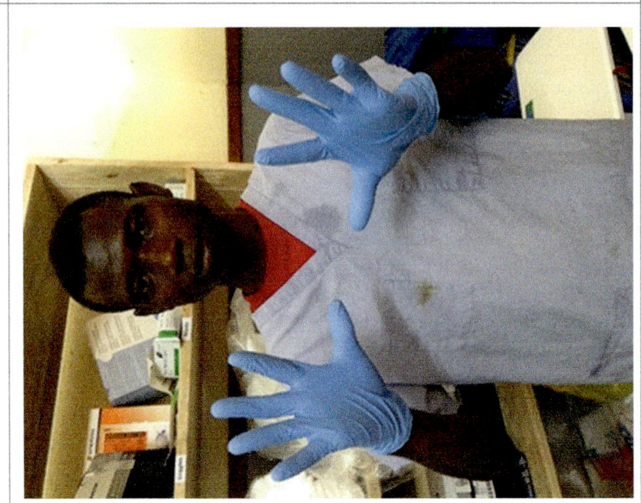

Step 13: Stand in chlorine bucket for 1 min and spray boots

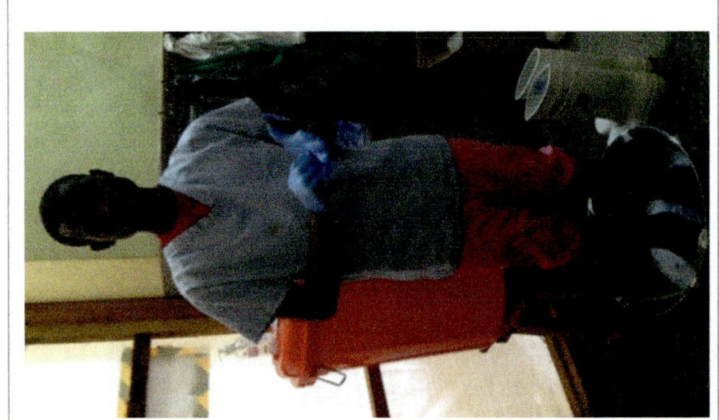
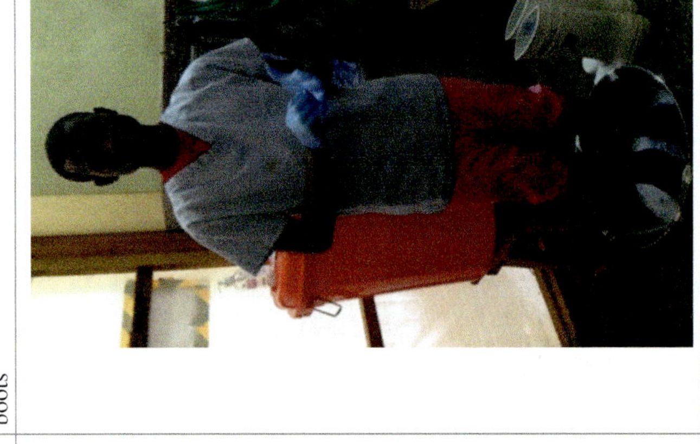

Step 14: Leave isolation unit through back door

Step 15: Remove gloves Wash hands with water and soap up to elbows (1/100)	**Step 16**: Put on gloves	**Step 17**: Enter dressing room though dressing room doors
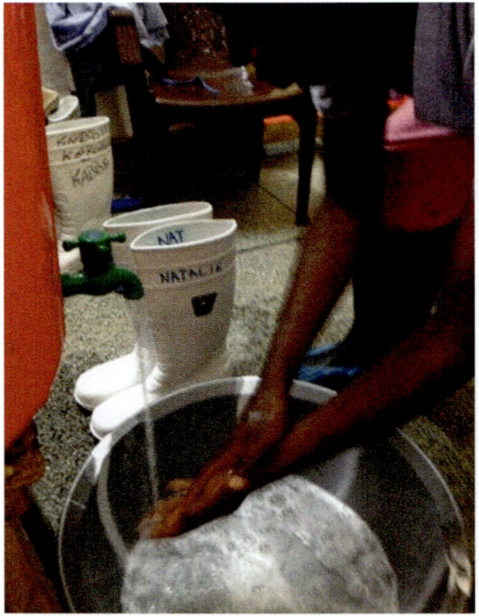	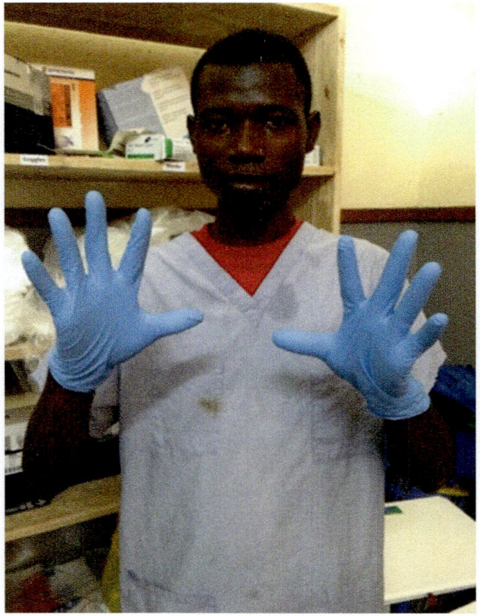	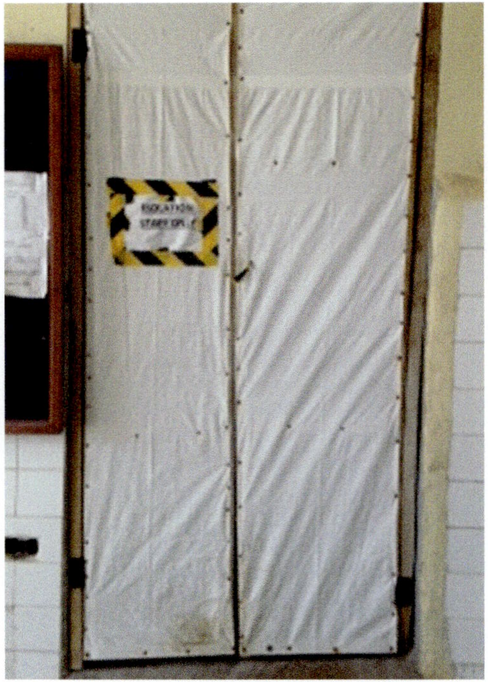

(continued)

Step 18: Remove boots Put on shoes	**Step 19**: Wash hands to elbows soap + water

Leave dressing room Remove shoe covers of normal shoes, wash hands with alcohol gel

3.2.5   Cleaning and Waste Management

Floors and horizontal work surfaces should be cleaned twice per shift by the cleaning team. Cleaning should always be carried out from "clean" areas to "dirty" areas, in order to avoid contaminant transfer. Any body fluids on surfaces (vomit, blood etc.) should be cleaned immediately. All soiled linen should be removed and placed in waste bins, never shake out soiled linen. Disinfection inside the unit should be done using 0.5% chlorine solution.

3.2.6   Safe Cleaning of Spills

Deal with any spills (vomit, blood, excreta or urine) **immediately** as they might be stepped in or slipped on.

- Pour 0.5% solution around the spill and then over it so that it is completely covered and leave it for 15 min by the clock. Take care to avoid any splashes.
- Use adsorbent material or draw sheets to mop up the spill and then put them in a waste bucket.
- Repeat the procedure until the area is clean.
- Apply final disinfection with 0.5%

Remember to pour the chlorine on to a spill rather than spray it to avoid aerosolisation of infectious material

3.2.7   Preparing Chlorine Solutions

There are two chlorine solutions used inside the isolation unit.

1. 0.5% solution (1 in 10) is used for:
    - Disinfection of body fluids
    - Clean up of spills
    - Disinfection of corpses
    - Disinfection of gloved hands
    - Disinfection of boots and aprons
    - In footbaths
    - Disinfection of floors, structures and furniture
    - Disinfection of beds and mattress covers

2. 0.05% solution (1 in 100) is used for:
   - Disinfection of bare hands and skin.
   - Sponge baths for patients

The cleaners and some nurses are responsible for making chlorine solution inside the unit. This happens three times a day at the beginning of each shift.

Standard chlorine is made from dry chlorine granules of calcium hypochlorite.

To make chlorine solution:

Volume (L)	0.5% (strong)	0.05% (weak)
1	7.5 g	1 part 0.5% with 9 parts water
10	75 g (5 heaped tbsp)	7.5 g
20	150 g (10 heaped tbsp)	15 g
60	450 g (30 heaped tbsp)	45 g

Practically, there is always a large bucket of 0.5% chlorine in the corner of the cleaners room inside the unit. This can be used to make 0.05% chlorine using the rule of "1+9":

0.05% (1 in 100)

- **1** (one) measure of the just prepared 0.5% solution plus **9** (nine) measures of clean water = a 0.05% solution

Chlorine solutions loose potency over time and in the sun, thus they must be prepared every few hours

Diagram, explaining the 1 in 10 rule

(note—this refers to using liquid 5% bleach initially):

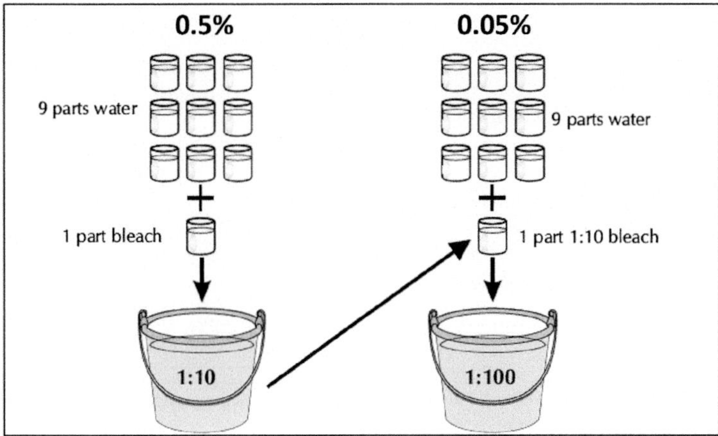

How to Make Chlorine Solutions

1. Solutions lose their effectiveness after 24 h, so they must be refreshed every day. Organic material (e.g., vomitus, dirt, etc.) also binds chlorine, decreasing its effectiveness; therefore, change disinfectant solutions whenever they become cloudy or visible dirty.
2. Always check the concentration and expiration date of chlorine before making chlorine solutions. Unknown concentrations and expired chlorine should not be used.
3. A simple suggestion to ensure that chlorine solutions are correctly made up each day is to mix the solution in one very large bucket/tank inside the Red Zone that can be used to fill the smaller buckets. The total volume of water required to fill this large bucket (either to the top or a clearly visible line) can measured by counting the number of litres of water this requires. The exact volume of chlorine powder required to make it to a 0.5% concentration can then be calculated (using high-accuracy weighing scales and the formulas below) and a plastic cut can be cut to the correct size so that one single full scoop of chlorine will always be the correct amount of chlorine for a single full large bucket of water. It can be helpful to buy 100 L buckets for each of the units to make this easier.

Colour coordination of different buckets depending on their concentration is an intuitive way to help staff identify which bucket to use for different tasks.

70% HTH chlorine powder:	*5% Liquid Bleach*:

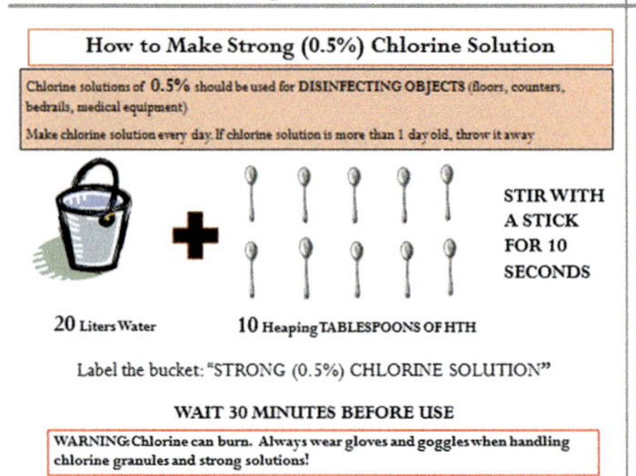

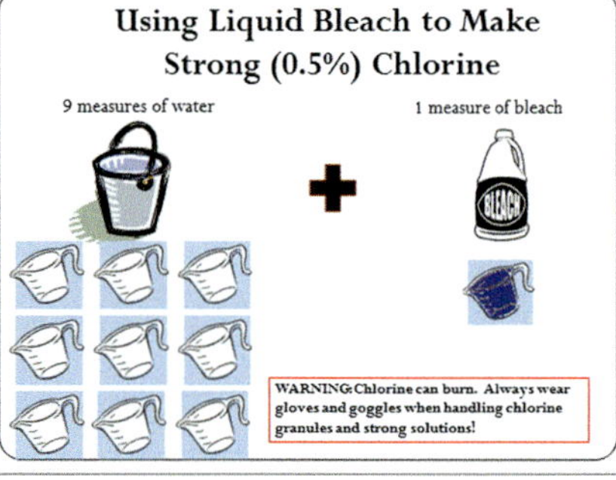

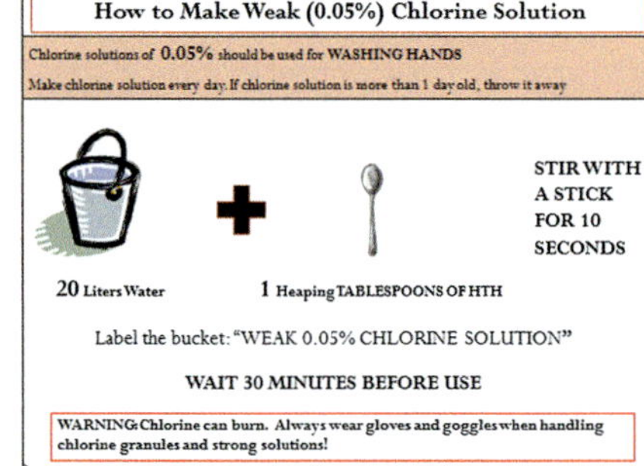

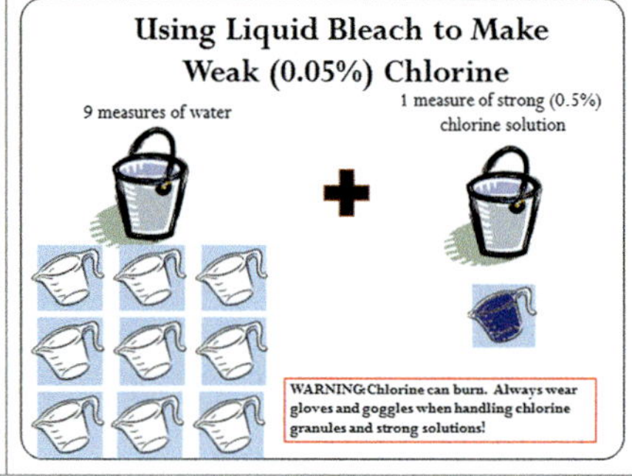

3.5% Liquid Bleach:

Strong (0.5%) chlorine solution:

Dilute 1 measure/part (1Litre) of 3.5% liquid bleach with 6 measures/parts (6 L) of water.

Weak (0.05%) chlorine solution:

Dilute 1 measure/part (1 L) of strong (0.5%) chlorine solution with 9 measures/parts (9 L) of water.

Any Other Chlorine Concentration:

Example I - Using Liquid Bleach

Chlorine in liquid bleach comes in different concentrations. Any concentration can be used to make a dilute chlorine solution by applying the following formula:

$$\left[\frac{\% \text{ chlorine in liquid bleach}}{\% \text{ chlorine desired}} \right] - 1 = \text{Total parts of water for each part bleach} \dagger$$

Example: To make a 0.5% chlorine solution from 3.5%‡ bleach:

$$\left[\frac{3.5\%}{0.5\%} \right] - 1 = 7 - 1 = 6 \text{ parts water for each part bleach}$$

Therefore, you must add 1 part 3.5% bleach to 6 parts water to make a 0.5% chlorine solution.

†   "Parts" can be used for any unit of measure (e.g. ounce, litre or gallon) or any container used for measuring, such as a pitcher.

‡   In countries where French products are available, the amount of active chlorine is usually expressed in degrees chlorum. One degree chlorum is equivalent to 0.3% active chlorine.

Example II - Using Bleach Powder

If using bleach powder,† calculate the amount of bleach to be mixed with each litre of water by using the following formula:

$$\left[\frac{\% \text{ chlorine desired}}{\% \text{ chlorine in bleach powder}} \right] \times 1\,000 = \text{Grams of bleach powder for each litre of water}$$

Example: To make a 0.5% chlorine solution from calcium hypochlorite (bleach) powder containing 35% active chlorine:

$$\left[\frac{0.5\%}{35\%} \right] \times 1\,000 = 0.0143 \times 1\,000 = 14.3$$

Therefore, you must dissolve 14.3 grams of calcium hypochlorite (bleach) powder in each litre of water used to make a 0.5% chlorine solution.

†   When bleach powder is used; the resulting chlorine solution is likely to be cloudy (milky).

Example III - Formula for Making a Dilute Solution from a Concentrated Solution

$$\text{Total Parts (TP) (H}_2\text{O)} \quad = \left[\frac{\% \text{ Concentrate}}{\% \text{ Dilute}} \right] - 1$$

Example: To make a dilute solution (0.1%) from 5% concentrated solution.

$$\text{Calculate TP (H}_2\text{O)} \quad = \left[\frac{5.0\%}{0.1\%} \right] - 1 = 50 - 1 = 49$$

Take 1 part concentrated solution and add to 49 parts boiled (filtered if necessary) water.

Source:
AVSC International (1999). Infection Prevention Curriculum. Teacher's Manual. New York, p.267.

3.2.8   Sharps Management

In order to prevent injuries from needles and other sharp instruments, use care when handling, using, and disposing of needles and medication vials. Please adhere to the following:

- When opening medication vials, use the plastic vial openers provided
- Do not recap needles
- Take the sharps container to the bedside when giving injections. Discard single-use needles and syringes immediately after use and directly into the sharps container, without recapping and without passing to another person
- Close, seal, and send sharps containers for incineration before they are completely full.

3.2.9   Accidental Exposure

Immediate Steps for Accidental Exposures Inside the Red Zone

Needle stick injury or cut with glass vial

1. Get a colleague to get a fresh bucket of 0.5% (1 in 10) chlorine
2. Immediately immerse the site in 0.5% chlorine solution for 3 min or until the site stops bleeding
3. Decontaminate as per protocol and leave isolation unit

4. Flush site with running water for 1 min
5. Apply dressing
6. Inform clinical lead

Contact with infectious bodily fluids (splash in eye)

1. Call a colleague for assistance
2. Colleague to wash hands in 0.5% chlorine (1 in 10), put on a fresh pair of gloves and wash hands again in 0.5% chlorine (1 in 10)
3. Use saline vials (available in the stock room) and get a colleague to rinse out eye
4. Decontaminate as per protocol and leave isolation unit
5. Leave isolation unit and remove PPE
6. Rinse out eye again
7. Inform clinical lead

Contact with infectious bodily fluids with broken skin

1. Wash the area with 0.5% (1 in 10) chlorine solution
2. Decontaminate as per protocol and leave isolation unit
3. Inform clinical lead

3.3   Management of Corpses

All corpses of Ebola positive and unknown status will be taken for unmarked burial. Corpses are stored in a dedicated room inside the isolation unit, and collected daily by the burials team.

A. Management of corpse (inside the isolation unit)

1. Check if a blood sample for Ebola has been taken, if not, an oral swab* needs to be taken

2. Dress in full PPE, enter the unit and proceed to patient bed

3. Assess floor safety, faeces, vomitus, blood etc.

4. Visually assess for signs of life. Feel for pulse and signs of breathing

(continued)

5. Check wristband to ensure correct patient

↓

6. Take oral swab* if needed

↓

7. Clean body with 0.5% chlorine (1 in 10) using a pan and bucket

↓

8. Prepare body bag and tag (all body bag tags should contain patient name and ID number)

↓

9. Roll or lift patient in to body bag. NEVER ALONE. Zip body bag close

↓

10. Wash body bag with 0.5% chlorine (1 in 10)

↓

11. Transfer body bag to the store room by the dirty exit

↓

12. Wash gloved hands in 1/10 chlorine

↓

13. Clean patient area and dispose of possessions

B. Management of corpse (outside the isolation unit)

1. Dress in full PPE, and approach the area

↓

2. Assess floor safety, faeces, vomitus, blood etc.

↓

3. Visually assess for signs of life. Feel for pulse and signs of breathing

↓

4. Spray down the body before handling with 0.5% chlorine (1 in 10)

↓

5. Transfer body onto stretcher and bring inside the isolation unit

↓

6. Take oral swab*

↓

7. Clean body with 0.5% chlorine (1 in 10) chlorine

↓

8. Prepare body bag and tag (all body bag tags should contain patient number and ID number)

↓

9. Roll or lift patient in to body bag. NEVER ALONE. Zip body bag

↓

10. Wash body bag with 0.5% chlorine (1 in 10)

↓

11. Transfer body bag to the store room by the dirty exit

↓

12. Wash gloved hands in 1/10 chlorine

↓

13. Decontaminate stretcher and ensure area where body was taken from is decontaminated

3.4   Safe Medical Burials

Transmission of Ebola is related to close contact with a person during the most severe stages of acute illness, including after death. Ebola virus disease is a disease of social intimacy as the main infection pathways are through nursing of the sick and through preparation of corpses for burial.

Funeral related events are a well-recognized opportunity for infection transmission. Viable virus has been isolated from animal tissues or fluids in the laboratory as late as 7 days post-mortem. However, not all the studies found such an association between attending funerals and disease risk. Furthermore, even for those attending funerals were transmission occurs, only people who engaged in certain behaviors are particularly at risk.

Attendance at a safe burial (Ebola corpse in body bag and buried according to protocol) or attendance without contact with the body and without contact with those who touched the body or any fluids or materials that contacted the body are not considered contacts.

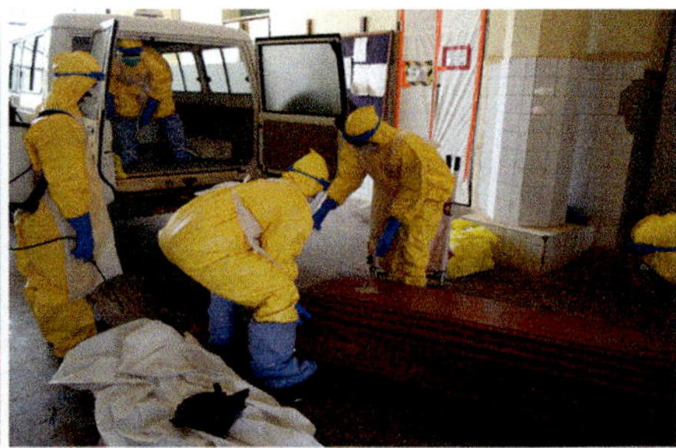

*Burial team in PPE managing the transportation of a corps for his burial

How to Prepare a Dead Body for a Safe Medical Burial
Burial should take place as soon as possible after the body is prepared in the health facility. Health facility staff should:

- Prepare the body safely.
- Be aware of the family's cultural practices and religious beliefs. Help the family understand why some practices cannot be done because they place the family or others at risk for exposure.

- Counsel the family about why special steps need to be taken to protect the family and community from illness. If the body is prepared without giving information and support to the family and the community, they may not want to bring other family members to the health facility in the future. They may think that if the patient dies, the body will not be returned to them.
- Identify a family member who has influence with the rest of the family and who can make sure family members avoid dangerous practices such as washing or touching the body.

To prepare the body in the health facility:

1. Wear protective clothing as recommended for staff in the patient isolation area. Use thick rubber gloves as the second pair (or outer layer) of gloves.
2. Wash the body and the area around it with 0.5% chlorine solution.
3. Place the body in a "body bag" (mortuary sack) and close it securely. Wash the body bag with 0.5% chlorine solution.
4. If body bags are not available, wrap the body in two thickness of cotton cloth and soak with 0.5% chlorine solution. Then wrap the body in plastic sheeting. Seal the wrapping with plastic tape. Wash the body bag as in Step 3. Place the body in a coffin if one is available.
5. Transport the body to the burial site as soon as possible.

To prepare the site for burial:

1. The grave should be at least 2 m deep.
2. Explain to the family that viewing the body is not possible. Help them to understand the reason for limiting the burial ceremony to family only.

*Burial sites in Sierra Leone

Some guidance is also now available in the form of a protocol from the World Health Organization covering safe and respectful burial in village conditions. The protocol is a great improvement, since it now specifies that a pastor or imam be present in all Ebola funerals, and the involvement, at a distance, of family witnesses. Further flexibility will be encouraged if villagers participate in focus sessions where they have an active hand in agreeing on the safest compromises. Corpse washing should be discouraged, but if it cannot be avoided then it should be done only with biohazard protection. It is possible for dignified funerals to be held without high risk to those attending.

*Burial team in Sierra Leone praying with a Safe medical burial following the protocols

Appendix 1: Sample Stock Card

Item:
Pack Size:

Date	Action	Number received	Number taken	Balance

Appendix 2: Estimation of Items for Ebola Holding Unit: Medical and Non Medical Consumables

Description	Item detail	QTY (for 10 patients)	Comments	Source
Cleaning equipment				
Mop		2		
Mop bucket		2		
Shovel		1		
Hard Broom/ Heavy duties broom		2	1 inside and 1 outside	
Sprayer—large	Backpack 12 L	2	1 inside unit, 1 outside	May be available via CMS
Squegee mop		2		

(continued)

Plasticware				
Buckets	20 L	7	For 1:10 chlorine for hand washing	
Buckets	10 L	12	10 for patients' washing water, 2 for cleaning	
Veronica buckets	20–30 L	3	Bucket with tap for 1:100 chlorine	
Jugs	1 L	3	For water filling	
Big bins	150 L	5	1 in dressing area, 1 in unit, 2 in decontamination area, 1 for storing 1:10	
Bowls	10 L	3	For decontamination area	
Furniture				
Shelves	2 shelves × 2 m	1	For supplies in dressing room	
Tables	1.2 × 0.6 m	1	1 in dressing area	
	2 × 0.6 m	2	1 in unit, decontamination area	
Bedside cabinet		10	for patient belongings	
Storage cabinet		1	Lockable with drawers, for drug storage	
Plastic chairs to be modified for toilet use		10		
Plastic chairs		2	For office use	
Furnishings				
Baby cot		1		
Plastic sheeting for screens between beds	1.25 m doubled			
String to hang sheeting				
Light bulbs		6		
Foot trays	1 m × 0.5m × 30 cm	2	To fill with chlorine and staff step into it to clean boots (plastic or metal)	
Barrels—metal	70 gallons	2	For burning waste	
Radio		1		
Mirror	30 cm wide	2	1 for dressing room and 1 for decontamination room	
Fan if electrical socket available		1		
Clocks		2		
Boot rack		1		
Boot puller		1		
Stretcher		1		

(continued)

Item	Unit	Quantity	Specification	Notes
Whiteboard and pens		1	Size depends on unit	
Uniform for health workers (Scrubs)				
Material (100% cotton) for × 30 scrubs (set)	76 yards	76		
Boots	Pair	10		Available via CMS

Medical Supplies (Estimate)

Item	Unit	Specification	Weekly estimate for 10 beds	Cost (£)	Notes
PPE suits	Single	XL or L	350	3.84	
Aprons	Single		400	0.04	
Hairnets or surgical caps	Single		350	From CMS	
Body bags	Single	Adult	15	6.34	
		Child	5	From CMS	
Chlorine powder		Ensure % concentration is stated on barrel	25 kg	From CMS	
Blood sample tubes	Single	Red top tubes	70	From CMS	
Plastic hazard waste bags for blood samples	Single	50 × 50 cm	70	From CMS	
Alcohol gel	500 ml bottle	Must be at least 70% and clearly marked	21	3.94	
Shoe covers	Single		1000	0.13	Per single cover
Masks with face shields	Single		400	0.26	
Gloves	Single	Nitrile gloves are preferable	3000	0.02	
Long gloves (useful for cleaners)	Single	Gynaecological gloves (40 cm length)	300	From CMS	Limited supplies
Rubber gloves—industrial	Pairs		10	From CMS	

Non-medical Consumables (Estimate)

DESCRIPTION	ITEM DETAIL	PER BOX	QTY for 1 week (13 patients)	QTY for 1 week (6 patients)	UNIT (SLL)	Total (SLL)
Patient drinks and food						
Juice (Tetra Pack) - Box of 27	200 ml	1	Err:522	1.50	40,000.00	60,000.00
Baby formula	400 g	1	0	2.00	25,000.00	50,000.00
Bottle for formula milk						0.00
Packet water	Bundle	1	50	25.00	3,000.00	75,000.00
Bottled water						
Disposable plates				150.00		
Disposable cups				150.00		
Disposable spoons				150.00		
Patient hygiene						
Toothpaste - Maxam	45 g	1	34	16.00	1,500.00	24,000.00
Toothbrush	Unit	1	34	16.00	1,000.00	16,000.00
Pillows	Unit	1	25	16.00	12,500.00	200,000.00
Bedsheets (fitted)	Unit	1	25	16.00	20,000.00	320,000.00
Pillowcases	Unit	1	25	16.00	10,000.00	160,000.00
Towels (medium size -hygien use)	Unit	1	17	8.00	12,000.00	96,000.00
Plastic Bowl for drinking water (4 packets of water)	Unit	1	10	5.00	4,000.00	20,000.00
Cleaning wipes for patients	Pack of 80	1	25	16.00	6,000.00	96,000.00
Pampers	Pack of 25	1	0	2.00	40,000.00	80,000.00
Cleaning materials						
Waste disposal bags (40 Gallons) - for big bins	Roll of 5	1	40	20.00	12,000.00	240,000.00
Waste disposal bags (25 litres) - to line toilet buckets	Roll of 10	1	10	5.00	8,000.00	40,000.00
Alcohol Hand Sanitizer (Bactigel) - 70% alcohol	350 ml	1	14	7.00	28,000.00	196,000.00
Cleaning paper Trendy (packet)	Pack of 100	1	3	2.00	10,000.00	20,000.00
Big roll (white/Blue) cleaning paper	Unit	1	3	1.00	20,000.00	20,000.00
Mopheads	Unit	1	7	3.00	10,000.00	30,000.00
Brush head	Unit	1		1.00	12,000.00	12,000.00
Cleaning Household Gloves	Pairs	1	0	7.00	8,000.00	56,000.00
Small black bin (5L) + Lid - toilet bucket	5L	1	30	15.00	5,000.00	75,000.00
Laundry Liquid Soap (Detoil/Rexoguard)	5L	1	12	6.00	14,000.00	84,000.00
Laundry Powder Soap (for Scrubs) - Foam	300-400 g	1	5	2.00	4,000.00	8,000.00
Toilet Paper	Units	1	5	3.00	2,500.00	7,500.00
Small towels for cleaning blood	Bundle	1	1.3	0.50	90,000.00	45,000.00
Kerosene for burning waste	Litres	1	28	14.00	4,500.00	63,000.00
Stationery						
Ledgers		1		2.00	15,000.00	30,000.00
Patient notes				0.00		0.00
Pens		1		10.00	500.00	5,000.00
Scissors		1		2.00	5,000.00	10,000.00
Admission forms						
Discharge negative forms						
Referral forms						
Patient clothes for 24 discharged patients						-
Lapa for woman (no dress)	Unit	1	0	12.00	10,000.00	120,000.00
T.Shirt (unisex) - Adults & kids	Unit	1	0	12.00	5,000.00	60,000.00
Trousers for men (Jeans, ...)	Unit	1	0	12.00	15,000.00	180,000.00
Trousers for kids & teenagers (unisex)	Unit	1	0	12.00	10,000.00	120,000.00
Underwear: men & boys (boxer, pants)	Unit	1	0	12.00	10,000.00	120,000.00
Underwear: women & girls (pants)	Unit	1	0	12.00	8,000.00	96,000.00
Flipflops	Pairs	1	0	24.00	5,000.00	120,000.00
Baby clothes						
TOTAL						2,954,500.00

Unit Stock Lists
Unit

Stock item	Stock level
Pillows	10
Bed sheets	20
T-shirts	6
Lappa	4
Trousers/shorts for men	4
Children's clothes	4 sets
Bin bags for black rubber bins	25
Towels for washing	4
Small towels for cleaning	10
Toothpaste	10
Toothbrush	10
Black rubber bins with lids	20
Green bowls	
Cleaning wipes	2 packets
Baby wipes	2 packets
Nappies	6
Juice	10 cartons
Packet water	120 packets
Bottle water	20 bottle

Drugs

Stock item	Stock level
Ceftriaxone 1 g injection	25
Artesunate 60 mg injection	20
ACT tablets adults	20 courses
ACT children 1–5 years	2 courses
ACT children 2–11 months	2 courses
Syringes 10 ml	20
Syringes 5 ml	40
Syringes 20 ml (for feeding)	10
Needles (blue)	100
Paracetamol 500 mg tablets	250
Paracetamol 125 mg/5 ml syrup	2 bottles
Metoclopramide 10 mg injection	20 vials
Metoclopramide 10 mg tablets	20 tablets
Tramadol 50 mg capsules	250
Diazepam 5 mg tablets	10 tablets
Diazepam 10 mg injection	4 vials

Unit Dressing Room

Stock item	Stock level
PPE—large	50
PPE—extra large	30
Masks with visors	100
Aprons	100
Sterile/long gloves—medium	2 boxes
Sterile/long gloves—large	2 boxes
Examination gloves—medium	4 boxes
Examination gloves—large	4 boxes
Shoe covers	300
Hair nets	50
Blood sample tubes	20
Biohazard bags for blood samples	20
Body bags—adult	10
Body bags—children	5
Surgical gowns	8
Hand soap	2 bottles
Hand sanitiser	2 bottles
Tissues	2 packets

Unit Office

Stock item	Stock level
Wristbands	20
Admission forms	20
DSO forms	20
Negative discharge forms	20
Survivor discharge forms	20
Referral forms	20
Dead positive forms	20
Dead negative forms	20
Drug charts	20
Hand soap	2 bottles
Hand sanitiser	2 bottles
Tissues	2 packets

Appendix 3: Examples of Order Forms

Date _______________   **Time** __________
Unit Store Room

Stock item	Stock level	Amount ordered
Pillows	10	
Bed sheets	20	
T-shirts	6	
Lappa	4	
Trousers/shorts for men	4	
Children's clothes	4 sets	
Bin bags for black rubber bins	25	
Towels for washing	4	
Small towels for cleaning	10	
Toothpaste	10	
Toothbrush	10	
Black rubber bins with lids	20	
Green bowls		
Cleaning wipes	2 packets	
Baby wipes	2 packets	
Nappies	6	
Juice	10 cartons	
Packet water	120 packets	
Bottle water	20 bottle	
Sharp bin	1	

Drugs

Stock item	Stock level	Amount ordered
Ceftriaxone 1 g injection	25	
Artesunate 60 mg injection	20	
ACT tablets adults	20 courses	
ACT children 1–5 years	2 courses	
ACT children 2–11 months	2 courses	
Syringes 10 ml	20	
Syringes 5 ml	40	
Syringes 20 ml (for feeding)	10	
Needles (blue)	100	
Paracetamol 500 mg tablets	250 tablets	
Paracetamol 125 mg/5 ml syrup	2 bottles	
Metoclopramide 10 mg injection	20 vials	
Metoclopramide 10 mg tablets	20 tablets	
Tramadol 50 mg capsules	250 capsules	
Diazepam 5 mg tablets	10 tablets	
Diazepam 10 mg injection	4 vials	

Unit Dressing Room

Stock item	Stock level	Amount ordered
PPE—large	50	
PPE—extra large	30	
Masks with visors	100	
Aprons	100	
Sterile gloves—medium	2 boxes	
Sterile gloves—large	2 boxes	
Examination gloves—medium	4 boxes	
Examination gloves—large	4 boxes	
Long gloves—medium		
Long gloves—large		
Long gloves—extra large		
Shoe covers	300	
Hair nets	50	
Blood sample tubes	20	
Biohazard bags for blood samples	20	
Body bags—adult	10	
Body bags—children	5	
Surgical gowns	8	
Hand soap	2 bottles	
Hand sanitiser	2 bottles	
Tissues	2 packets	

Unit Office

Stock item	Stock level	Amount ordered
Wristbands	20	
Admission forms	20	
DSO forms	50	
Negative discharge forms	20	
Survivor discharge forms	20	
Referral forms	20	
Dead positive forms	20	
Dead negative forms	20	
Screening forms	50	
Drug charts	20	
Hand soap	2 bottles	
Hand sanitiser	2 bottles	
Tissues	2 packets	

Ordered by: ____ **(print name)**

Name of facility	
Inspected by	
Current bed capacity	
Date of inspection	

Appendix 4: Evaluation of Isolation Unit Forms

Clinical Checklist	SCO	Good	Acceptable	Unacceptable	Comments and actions
Identify Unit Coordinator and Clinical Lead	O				
Identify Supervising Partner	O				
How many clinical staff working in unit? Local/expat	O				
How many non-clinical staff working in unit?	O				
Clinical Protocol pack displayed	O				
Ensure all staff entering the unit have undergone PPE training	S O				
Check staffing and attendance	O				
Any holding unit staff infections since last assessment?	S				
Has risk allowance been paid in last 2 weeks?	O				
Check registration and admission documents and DSO forms	O				
Check patient documentation	O				
Check for special populations: paediatric, pregnant, appropriate for unit?	OS				
Are stock levels of PPE above minimum stock levels? (see supply checklist)	S O				
Is stock accessible 24 hours a day	S O				
Is stock stored in safe, dry environment	S O				
Is physical security present and safe at facility?	S O				
Adequate water supply?	S O				
24-hour lighting/electricity/water available?	S O				
Screening, is screening area safe?	S				
Is screening tool being used?	O C				
No touch screening process and PPE?	S				
Dressing room are PPE protocols displayed	O S				
Observe staff putting on PPE, choose 2 at random	S				
Clear demarcation of Red Zone understood, no staff crossing line?	S				
Any PPE failings inside unit?	S				
Hand washing between contacts?	SC				
Appropriate amount of physical contact between staff and patients?	SC				
Respecting clean>dirty flow?	S				
Obvious unsafe practices?	S				
Correct bed to patient ratio	S				
Beds still set up to unit original design:	S				
Are chlorine footbaths full?					
Are chlorine hand washing buckets full?	S				
Is chlorine made up correctly? (test kit)	S				

Clinical Checklist	SCO	Good	Acceptable	Unacceptable	Comments and actions
Correct patient population to register (check wristbands)	O				
Correct patient bodily waste items (latrine chair/bucket)	C S				
Good isolation between beds	C S				
Do patients appear well cared for?	C				
Bedside area clean?	C				
Availability of ORS and water at patient bedside	C				
Hydration status of patients	C				
Directly question patients if they have received medication	C				
Directly ask if they have been receiving regular food	C				
Directly ask if patients have any specific concerns	C				
Directly take history from any patient who looks asymptomatic or other pathology suspected	C				
Observe unit Clinical Lead performing clinical review/ward round on patients.	C				
Observe drawing up/preparation of medication	C S				
Observe administration of medication	C S				
Observe proper sharps disposal and check sharps bins	C S				
Observe corpse removal if timing allows, if not have staff demonstrate knowledge	S				
Observe oral swab or blood sample collection if present	C S				
Check storage of blood specimens	S				
Check documentation of blood samples and timing	O				
Check proper storage of corpses at facility and timely burial	S				
Observe waste disposal	S				
Observe decontamination, choose 2 random staff	S				
All items necessary are in decon room?	S				
Overall assessment of organisation	O				
Overall assessment of clinical care given to patients	C				
Overall assessment of safety inside unit	S				

Domain: *S* Safety: staff and patient, *C* Clinical Care, *O* Organisation

*Unacceptable marks in the safety and clinical care domains should lead to immediate action being taken.

Actions

Appendix 5: Training Forms for Directly Observed Procedures for Clinical Staff

Hand washing	
6 steps observed	
Completed?	

Putting on PPE	
Identify correct equipment	
Put on gloves and foot covers	
Enter room	
Change into boots	
Put on suit	
Put on apron	
Put on second pair of gloves	
Put on mask	
Check in mirror	
Enter Unit	
Completed?	

Taking off PPE	
Phase 1	
Wash gloves for 1 min (1/10)	
Remove apron and put in bin	
Wash gloves for 1 min 1/10	
Remove apron and put in bin	
Wash gloves for 1 min 1/10	
Remove outer gloves	
Phase 2	
Wash gloves for 30 s 1/10	
Remove suit	
Wash gloves for 30 s 1/10	
Remove facemask	
Wash gloves 1/10	
Remove inner gloves	
Wash hands for 1 min up to elbows in 1/100	
Phase 3	
Put on gloves	
Stand in chlorine bucket 1 min	
Spray own boots	
Remove gloves	

(continued)

Wash hands 1/100	
Put on gloves	
Leave isolation unit through back door	
Enter dressing room through dressing room door	
Remove boots	
Put on shoes	
Remove shoe covers and gloves	
Wash hands Use alcohol gel	
Leave dressing room	
Completed?	

Making up chlorine	
60/70%–72 g per 10 L makes 10 L of 1:10	
Master mix bin must be filled whenever empty	
1 L of 1/10 +9 L normal water makes 10 L 1/100	
Change chlorine for fresh mix every 24 h	
Change foot baths every 12 h	
Completed?	

Dead body in the unit	
Put on PPE	
Identify patient and confirm death	
Wash chlorine	
Put body in bag (2/3 people)	
Attach label to body bag	
Move body bag to corridor	
Spray/wash previously occupied space	
Completed?	

Dead on Arrival	
Collect orange specimen pot form lab upstairs	
Put on PPE	
Identify patient and confirm death	
Do 2 oral swabs + place in same container	
Put specimen pot in corridor	
Wash body with chlorine	

(continued)

Put body in body bag (2/3 people)	
Attach label to body bag	
Move to corridor	
Spray/wash previous occupied space	
Inform burial team	
Completed?	

Forms to observe	
DSO	
Admission	
Drug charts	
Oral swab	
Referral form	
Dead- positive	
Dead—negative	
Outcome—negative	
Completed?	

Taking Blood	
Identify pending collections from board	
Write label	
Enter ISO + Take blood	
Put blood in corridor box	
Update board	
Update book	
Completed?	

Accidental Exposure Theory	
Wash site of exposure thoroughly	
Decontaminate	
Inform senior	
Document	
Completed?	

Appendix 6: Clinical Lead Training Reporting Form

Unit	
Name	
Position	
Contact number	
Years of experience	

Training

	Date	Comments
Basic PPE Training		
King's Training		
Refresher Training		

Directly Observed Procedures

	Date	Pass	Comments
PPE On			
Supervisor signature			
PPE Off			
Supervisor signature			
Hand washing 6 steps			
Supervisor signature			
Making up chlorine			
Supervisor signature			
Administration PO medication			
Supervisor signature			
Administration IM medication			
Supervisor signature			
Corpse procedure			
Supervisor signature			
Patient admission			
Supervisor signature			
Accidental exposure theory			
Supervisor signature			
Assessment and treatment of patient in pain			
Supervisor signature			
Assessment and treatment of patient with nausea/vomiting			
Supervisor signature			
Assessment and treatment of agitated patient			
Supervisor signature			
Paediatric patient			
Supervisor signature			

(continued)

Oral Swab			
Supervisor signature			
Command flow understanding			
Supervisor signature			

Safe to work (yes/no):   **Supervisor signature:**   **Date:**
Comments and areas of improvement:
Date of next assessment:

Appendix 7: Cleaner Training Reporting Form

Unit	
Name	
Position	
Contact number	
Years of experience	

Training

	Date	Comments
Basic PPE Training		
King's Training		
Refresher Training		

Directly Observed Procedures

	Date	Pass	Comments
PPE On			
Supervisor signature			
PPE Off			
Supervisor signature			
Hand washing 6 steps			
Supervisor signature			
Make Up Chlorine			
Supervisor signature			
Clean infected area			
Supervisor signature			
Safe Waste disposal			
Supervisor signature			
Prepare clean bed			
Supervisor signature			
Accidental exposure (theory)			
Supervisor signature			

Appendix 8: Nurse Training Reporting Form

Unit	
Name	
Position	
Contact number	
Years of experience	

Training

	Date	Comments
Basic PPE Training		
King's Training		
Refresher Training		

Directly Observed Procedures

	Date	Pass	Comments
PPE On			
Supervisor signature			
PPE Off			
Supervisor signature			
Hand washing 6 steps			
Supervisor signature			
Administration PO medication			
Supervisor signature			
Administration IM medication			
Supervisor signature			
Corpse procedure			
Supervisor signature			
Patient admission/discharge forms			
Supervisor signature			
Accidental exposure theory			
Supervisor signature			

Safe to work (yes/no):   **Supervisor signature:**   **Date:**

If no, further remedial action needed:

Date of next assessment:

Appendix 9: Laboratory Technician Training Reporting Form

Unit	
Name	
Position	
Contact number	
Years of experience	

Training

	Date	Comments
Basic PPE Training		
King's Training		
Refresher Training		

Directly Observed Procedures

	Date	Pass	Comments
PPE On			
Supervisor signature			
PPE Off			
Supervisor signature			
Hand washing 6 steps			
Supervisor signature			
Labelling of specimen + white board + book documentation			
Supervisor signature			
Specimen collection			
Supervisor signature			
Safe sharps disposal			
Supervisor signature			
Sample storage			
Supervisor signature			
Oral swab collection			
Supervisor signature			
Accidental exposure (theory)			
Supervisor signature			

Safe to work (yes/no):   **Supervisor signature:**   **Date:**

If no, further remedial action needed:

Date of next assessment:

Appendix 10: Admission Form in the Holding Unit

PATIENT NAME		BED NUMBER
LABORATORY ID NUMBER		
AGE		GENDER	
ADDRESS		
DISTRICT		
RISK: CONTACT OR TRAVEL		
PROFESSION		If **HCW**, please record and **inform CDC**
SYMPTOMS ON ADMISSION	☐ Fever ☐ Vomiting/nausea ☐ Vomiting with blood ☐ Diarrhoea ☐ Diarrhoea with blood ☐ Intense fatigue/weakness ☐ Anorexia/LOA ☐ Abdominal pain ☐ Muscle pain ☐ Joint pain ☐ Headache ☐ Cough	☐ Difficulty breathing ☐ Difficulty swallowing ☐ Sore throat ☐ Jaundice ☐ Conjunctivitis ☐ Skin rash ☐ Hiccups ☐ Pain behind eyes ☐ Coma/unconscious ☐ Confused/disorientated ☐ Other bleeding ☐ Other symptoms
OTHER BLEEDING OR SYMPTOMS		
DATE OF ONSET SYMPTOMS		
DATE FIRST PRESENTED		
SOURCE OF ADMISSION	☐ A&E-TRIAGE TENT ☐ DSO ☐ WARD	☐ OTHER HOLDING UNIT ☐ PRISION ☐ DEAD ON ARRIVAL
DATE OF ADMISSION		
LABORATORY SAMPLE	TAKEN	SENT	RESULT
OUTCOME	☐ DISCHARGE TO HOME ☐ DISCHARGE TO WARD	☐ REFFERED TO .... ☐ DEATH
DATE OF OUTCOME		

Pulse	beats/min		
Respiratory rate	breaths/min		
Mucous membranes	dry/moist		
Neuro (AVPU)	Yes (Y) No (N) Not done (ND)		
A: Alert?		**If A = Yes**	Walks unaided
			Walks with assistance
			Sits unaided
			Sits with assistance
			Drinks unaided
			Drinks with assistance
			Recalls name
V: Responds to voice?			
P: Responds to pain?			
U: Unresponsive			
Jaundice			
Conjunctivitis			
Other comments:			
Date	DD/MM/YY		
Time	hh:mm		
Completed by	initials		

Appendix 11: Certificate of EVD Negative Result in a Suspected Case for Discharge

<Hospital Name>
<District>, Sierra Leone

The patient ________________________________

Has been isolated in <Hospital Name> from _________ to _________

And a blood sample has been taken to perform a test for EBOLA with a NEGATIVE result.

The patient is discharged from the Ebola Holding Unit.

Date: _______________

<Doctors Name>
<Hospital Name>

Signature

Appendix 12: Referral form for EVD Positive Cases to a Treatment Centre

Name of holding centre			
Patient name			
Laboratory ID number			
Age			
Address			
Family or carer contact name		Family or carer contact number	
Date of onset symptoms			
Date of admission to isolation unit		Date of transfer to treatment centre	
Symptoms (tick relevant ones)	**General** ☐ Fever ☐ Weakness/severe fatigue ☐ Generalised body/joint pain ☐ Back pain **Gastrointestinal** ☐ Sore throat ☐ Difficulty swallowing ☐ Chest/epigastric pain ☐ Hiccups ☐ Abdominal pain ☐ Anorexia/loss of appetite ☐ Nausea ☐ Vomiting ☐ Diarrhoea	**Skin and mucoses** ☐ Skin rash ☐ Conjunctivitis ☐ Neurological status ☐ Confused/disorientated ☐ Coma/unconscious **Bleeding** ☐ Oral/gum bleeding ☐ Haemoptysis ☐ Haematemesis ☐ Melena ☐ Epistaxis ☐ Vaginal bleeding ☐ Haematuria ☐ Other site ______ **Other symptoms**	
Medication received in isolation unit			

A blood sample has been taken and this patient tested POSITIVE for EBOLA.
This patient is referred to your facility for further treatment and management.

Signed by:

Name: Position:

Appendix 13

Connaught Hospital Infectious Diseases Unit
PPE DONNING: FULL PPE

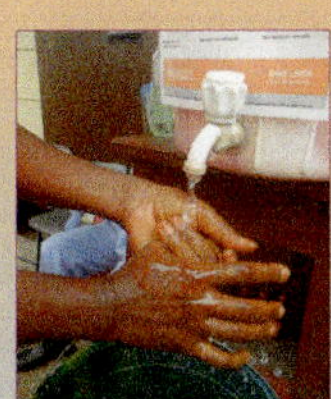

1. Wash hands before entering donning room

2. Put on one pair of examination gloves

3. Put on rubber boots

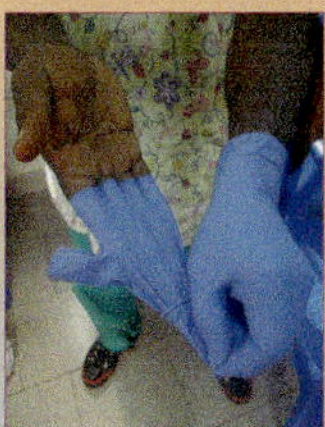

4. Remove gloves

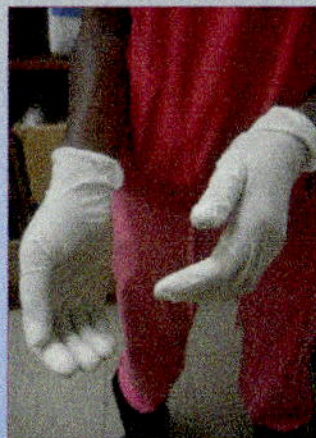

5. Put on one clean pair of examination gloves

6. Put on PPE suit, covering your boots to the ankles

7. Attach loops around thumbs

8. Put on an apron and tie it at the back

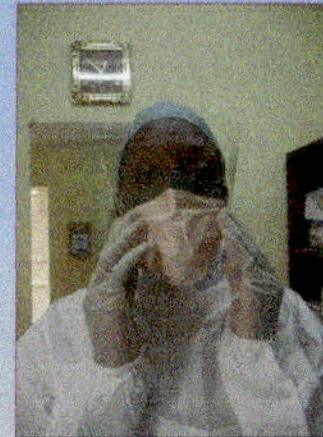

9. Put on face mask and eye cover

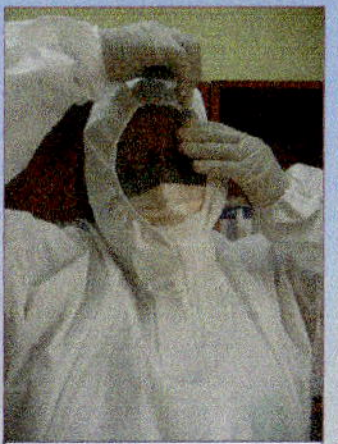

10. Fix hood over the top of mask and eye cover

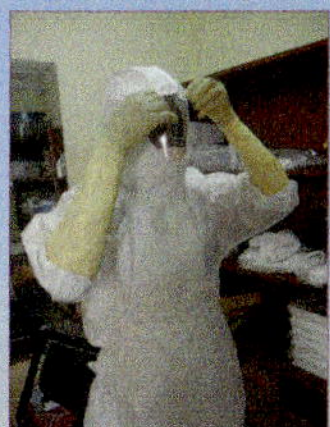

11. Put on one pair of elbow gloves

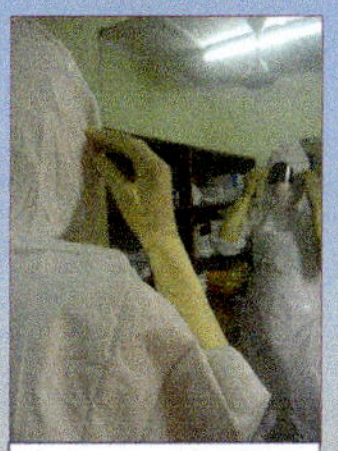

12. Check yourself in the mirror and adjust suit

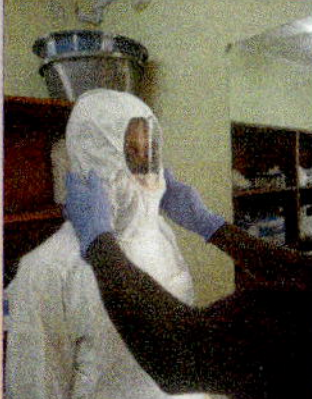

13. Get a colleague to check you before entering

REMEMBER:

✓ Ensure every part of your body is covered before entering the unit

✓ NEVER enter the unit alone, always ensure you have a colleague with you

✓ If you have any breaks in your skin, make sure you consult the unit manager before entering

✓ If your PPE gets damaged or dislodged while you are inside the unit, inform your colleagues who are in the unit with you and carefully go and decontaminate

✓ If you feel unwell while you are inside the unit, inform your colleagues who are in the unit with you and carefully go and decontaminate

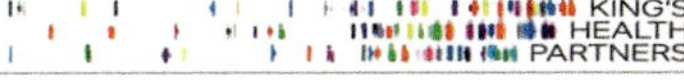

Appendix 14

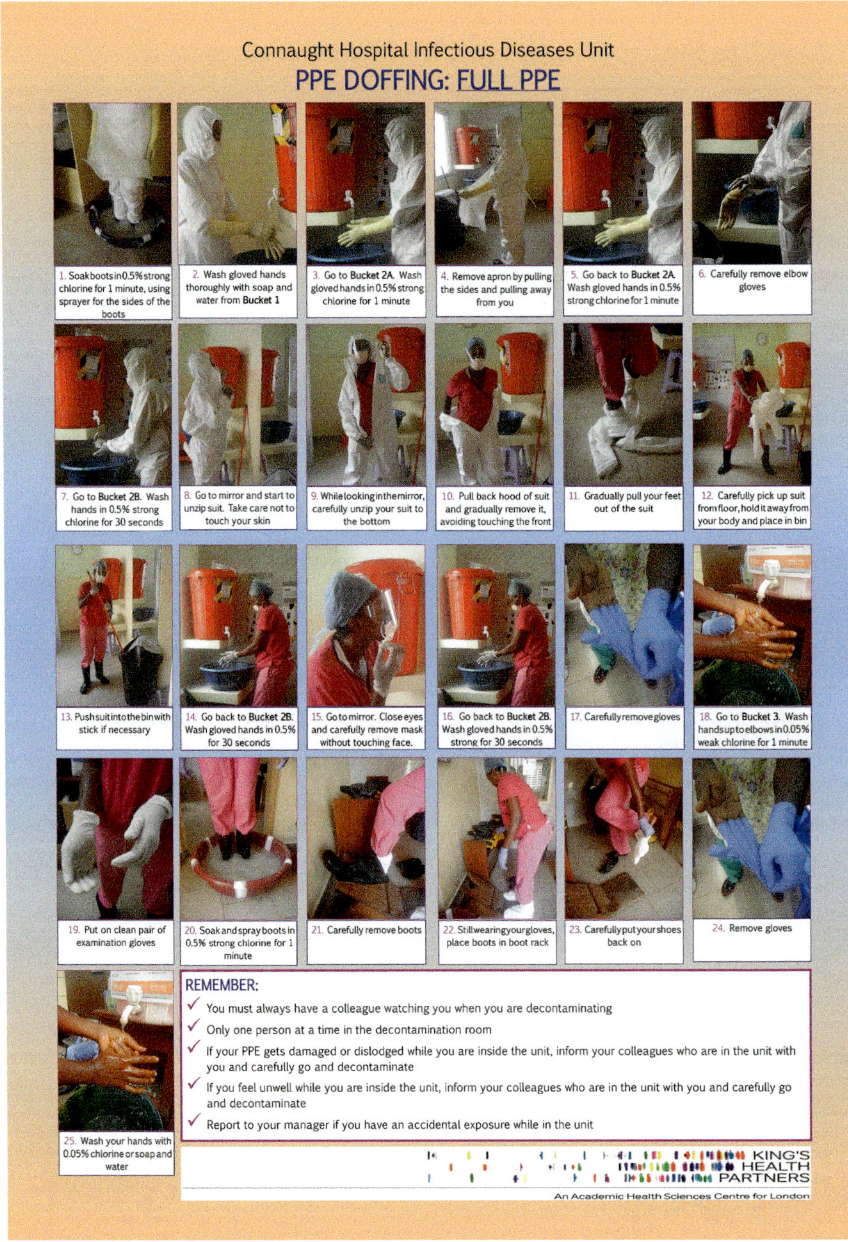

Connaught Hospital Infectious Diseases Unit
PPE DOFFING: FULL PPE
1. Soak boots in 0.5% strong chlorine for 1 minute, using sprayer for the sides of the boots
2. Wash gloved hands thoroughly with soap and water from Bucket 1
3. Go to Bucket 2A. Wash gloved hands in 0.5% strong chlorine for 1 minute
4. Remove apron by pulling the sides and pulling away from you
5. Go back to Bucket 2A. Wash gloved hands in 0.5% strong chlorine for 1 minute
6. Carefully remove elbow gloves
7. Go to Bucket 2B. Wash hands in 0.5% strong chlorine for 30 seconds
8. Go to mirror and start to unzip suit. Take care not to touch your skin
9. While looking in the mirror, carefully unzip your suit to the bottom
10. Pull back hood of suit and gradually remove it, avoiding touching the front
11. Gradually pull your feet out of the suit
12. Carefully pick up suit from floor, hold it away from your body and place in bin
13. Push suit into the bin with stick if necessary
14. Go back to Bucket 2B. Wash gloved hands in 0.5% for 30 seconds
15. Go to mirror. Close eyes and carefully remove mask without touching face.
16. Go back to Bucket 2B. Wash gloved hands in 0.5% strong for 30 seconds
17. Carefully remove gloves
18. Go to Bucket 3. Wash hands up to elbows in 0.05% weak chlorine for 1 minute
19. Put on clean pair of examination gloves
20. Soak and spray boots in 0.5% strong chlorine for 1 minute
21. Carefully remove boots
22. Still wearing your gloves, place boots in boot rack
23. Carefully put your shoes back on
24. Remove gloves
25. Wash your hands with 0.05% chlorine or soap and water
REMEMBER:
You must always have a colleague watching you when you are decontaminating
Only one person at a time in the decontamination room
If your PPE gets damaged or dislodged while you are inside the unit, inform your colleagues who are in the unit with you and carefully go and decontaminate
If you feel unwell while you are inside the unit, inform your colleagues who are in the unit with you and carefully go and decontaminate
Report to your manager if you have an accidental exposure while in the unit
KING'S HEALTH PARTNERS
An Academic Health Sciences Centre for London

Ebola Virus Disease in the Obstetric Population

4

Colin S. Brown, Diana Garde, Emily Headrick, Felicity Fitzgerald, Andy Hall, Hooi-Ling Harrison, and Naomi F. Walker

Contents

C. S. Brown
King's Sierra Leone Partnership, King's Centre for Global Health, King's Health Partners, King's College London, London, UK

Department of Infection, Royal Free London NHS Foundation Trust, London, UK

National Infection Service, Public Health England, London, UK

D. Garde (✉) · E. Headrick
Partners in Health (Sierra Leone, deployed 1/2015–7/2016), Boston, MA, USA

F. Fitzgerald (✉)
Infection, Immunity, Inflammation and Physiological Medicine, UCL Great Ormond Street Institute of Child Health, London, UK

A. Hall
King's Sierra Leone Partnership, King's Centre for Global Health, King's Health Partners, King's College London, London, UK

H.-L. Harrison
King's Sierra Leone Partnership, King's Centre for Global Health, Weston Education Centre, London, UK

N. F. Walker (✉)
Department of Clinical Research, London School of Hygiene and Tropical Medicine, London, UK

Hospital for Tropical Diseases, University College London Hospital, London, UK
e-mail: naomi.walker@lshtm.ac.uk

© Springer International Publishing AG, part of Springer Nature 2018
M. Lado (ed.), *Ebola Virus Disease*, https://doi.org/10.1007/978-3-319-94854-6_4

4.1 Management of Ebola Virus Disease in the Low and High Resource Setting

Naomi F. Walker, Colin S. Brown, and Hooi-Ling Harrison

4.1.1 Introduction

The clinical management of Ebola created a significant challenge during the outbreak in West Africa in 2014/15, due to the paucity of previous research conducted into the optimum treatment regimen. That left many centres, to some extent, having to 'work out' best practice as they went along, and attempting to conduct real time prospective research. Médecins Sans Frontières (MSF) [1] were the only organization to have provided relatively in depth practical guidance prior to the outbreak and this manual was the basis of further planning between the WHO, national Ministry of Health and Sanitation in Sierra Leone, and other relevant stakeholders. Additionally, guidance changed over the epidemic as experience grew. This chapter will describe four key areas in the management of Ebola in West Africa. Firstly, it outlines the most recent WHO guidance; secondly, it looks back at how Ebola was managed in differing low and high resource settings; thirdly it outlines possible and optimal options for managing complications, paying particular attention to some of the controversies faced; fourthly it describes recent and ongoing studies into potential novel therapies that may shape future practice.

There is, as yet, no specific cure for EVD and the mainstay of treatment is supportive care and managing complications, alongside ring-fence vaccination of close contacts. However, due to many tropical illnesses masquerading as Ebola, and

Table 4.1 EVD differential diagnosis[a]

Differential diagnosis (consider local epidemiology)	Specific interventions
Typhoid	Antibiotics
Malaria	Malaria RDT, antimalarials
Viral/bacterial gastroenteritis e.g. *Shigella* spp. infection	ORS, IV hydration Consider antibiotics
Cholera	ORS
Amoebic dysentery	Anti-protozoal therapy e.g. metronidazole/tinidazole
Dengue fever	Monitor for shock syndrome
Other VHFs e.g. Lassa, CCHF	Supportive, consider Ribavirin if confirmed Lassa/CCHF
HIV-related illness	VCT, further investigations when EVD excluded. Consider empirical therapy for sepsis/PCP depending on presentation
Leptospirosis	Antibiotics
Typhus/rickettsial disease	Antibiotics e.g. doxycycline
Other viral illness e.g. influenza, EBV	N/A
Other bacterial sepsis	Antibiotics, monitor and treat shock if present, inotropic support if required and available, investigate source

[a]Not exhaustive
N/A not applicable, *ORS* oral rehydration solution, *PCP* Pneumocystis pneumonia, *VCT* voluntary counselling and testing for HIV

thus requiring isolation and EVD testing, it is worth bearing these diagnoses in mind as empirical treatment for them should be considered in the undifferentiated patient (Table 4.1). This is particularly important in contexts lacking diagnostic resources, and additionally, with the current absence of a rapid diagnostic test that has been widely used or field tested. As an example, O'Shea et al. [2] reviewed 51 patients who presented with suspected EVD but tested negative, at the UK Ministry of Defense ETU at Kerrytown, Sierra Leone. Eleven had malaria (25.5%), 12 *Shigella* spp. or *E. coli*, 12 a respiratory infection and 11, an undifferentiated febrile illness supporting the need for empirical malaria and antibiotic treatment.

4.1.2 WHO Guidance 2016 [3]

The WHO released their most recent guidance on the management of Ebola in February 2016. This is recognised as the gold standard in the low-income setting, however, it is possible that the capabilities of many centres will not meet these targets. The WHO classifies the patient into three categories of severity; mild, moderate and severe. These are briefly summarised here, but for specific guidance the most up to date version of the WHO manual should be referred to.

Mild

The patient will be haemodynamically normal, able to eat and drink, have no complications nor evidence of dehydration. Empirical oral broad-spectrum antibiotics, anti-malarials and oral rehydration solution are recommended. Treatment for symptoms such as pain, fever and dyspepsia can be provided orally.

Moderate

Patients often have vomiting and diarrhoea, are weak and look dehydrated with sunken eyes and a skin pinch slow to return. Antibiotics and anti-malarials should ideally be given intravenously, however, if there are contraindications to intravenous line insertion, can be administered intramuscularly. Fluids should be given intravenously with additional oral rehydration solution.

Severe

This patient will be severely dehydrated and shocked from either dehydration, sepsis or bleeding, although haemorrhage in the West African outbreak was seen in as little as 5–10% of cases. Empirical broad-spectrum antibiotics should be administered intravenously within the first hour of presentation, along with anti-malarials, and effective intravenous fluid resuscitation is essential. Tables are provided in the WHO guidelines to calculate fluid requirements, however, as a rule of thumb an adult who is severely dehydrated or shocked would be treated with a bolus of 1L in the first 30 min. Monitoring the patient for acute fluid overload with regular vital sign measurement and fluid balance charts is required.

In ill patients, the WHO advocates testing for malaria and HIV, and measuring potassium, sodium, bicarbonate, blood glucose, creatinine, and lactate. For women of a childbearing age, a pregnancy test should be performed as a priority. If possible, magnesium, haemoglobin or hematocrit, platelet count, INR and APTT should also be tested.

In all instances management of symptoms and signs is required—these are described in the management of complications section below.

4.1.3 Practical Aspects of EVD Management in Differing Contexts

There was great variation in EVD management in practice, influenced in part by resource discrepancies, differences in case load and staffing, and disagreements in best practice that developed due to the void of evidence. Here we discuss some of the different EVD management settings, the type of care that was provided and the impact that these differences may have had on patients' outcomes.

Table 4.2 describes the variation in care across the three most affected West African countries at low and high resource centres, and those repatriated or treated within Europe and the United States. There was a wide variation in reported mortality rates. However, the heterogeneity of populations studied, demographics, geography, time period (early versus late in the outbreak), survivor bias (only

Table 4.2 Examples of management in Ebola care facilities—West African Ebola outbreak

Location/phase of outbreak	EHU/ETC/ECC	Organization	Patient characteristics	Resources/management strategy	References
Kenema, Sierra Leone EARLY Rural	Government hospital, EHU, ETC	MoH WHO	Local walk in and regional referrals. One of few ETCs available early in the outbreak, high burden area. Severe cases. High mortality (74%)	Low. Haematology and biochemistry possible. IV fluids for all patients. Antibiotics for most (93%). Anti-malarials for 55%	Schieffelin et al. [4]
Kailahun Sierra Leone EARLY Rural	EHU, ETC	MoH	Local walk in and regional referrals. Often presented late or already died in transport due to distance. Only 1 ambulance for three million population. 53% mortality	Low. No labs or IV fluids. Supportive and symptomatic care—ORS, antibiotics, anti-pyretics	Dallatomasina et al. [5]
Freetown Sierra Leone EARLY Urban	Government hospital, EHU	MoH KSLP	Local walk in and referrals. High burden area, patients with comorbidities. All confirmed cases referred to ETCs. Overall mortality—39% (unpublished data)	Low. IV fluids in some cases. Antibiotics and anti-malarials for all. Only HIV and malaria RDT possible, later in the outbreak	Lado et al. [6]
Conakry Guinea EARLY Urban	Two sites: ETC and government hospital	WHO MoH MSF	Walk ins and referrals. First ETC of outbreak. High burden. Mortality 43%	Low. POC labs for some. WHO/MSF protocolised care but poorly staffed limiting interventions. 72% IV fluids (average 1 L/24 h), all antibiotics, 18% anti-malarials	Bah et al. [7]
Hastings Sierra Leone MID Peri-urban	Military, ETC	MoH MoD (SL)	Referrals, confirmed cases only. Lower mortality (32%) reported than some concurrent studies but may be related to selection of patients (possible survivor bias). High burden area	Low. No labs. HIV and malaria RDT possible. Strict 72 h regime of IV fluids (500 mL R/L, 500 mL 5% dextrose alternating every 8 h, IV Antibiotics, IV anti-malarials, then switch to oral. NSAIDs, multivitamins, ORS and fruit juice	Ansumama et al. [8]

(continued)

Table 4.2 (continued)

Location/phase of outbreak	EHU/ETC/ECC	Organization	Patient characteristics	Resources/management strategy	References
Monrovia Liberia MID Urban	ETC	MSF	Walk ins and referrals. High burden	Low. No Labs. High patient: physician ratio (30–50:1) limited IV fluid use. Treated according to staging: clinically hypovolaemic but not in shock and self caring—antiemetic, anti-diarrhoeal, ORS; hypovolaemic, shocked and unable to self care—IV fluids; shock and organ dysfunction whose outcome would not be altered by medical intervention—no treatment	Chertow et al. [9]
Waterloo Sierra Leone MID Peri-urban	ETU	Chinese military	Referred and confirmed cases Mortality 69%, possibly from late presentations	Low. No labs. Protocolised WHO treatment given according to clinical judgement: anti-malarial, IV fluids, Antibiotics. High patient:staff ratio	Qin et al. [10]
KerrytownSierra Leone LATE Rural	EHU, ETC	Save the Children DFID	Mostly referral patients, confirmed cases. Lower mortality (37%), may be related to later phase of outbreak, patients referred earlier or survivor bias, better staff:patient ratio and improved monitoring	Medium: access to laboratory monitoring and relatively well-staffed. Treated according to staging of patient (decided by WHO and UK-MoD). Stage 1: ORS, multivitamins, antimalarial if RDT +ve, targeted electrolyte and glucose therapy. Stage 2: IV fluids 3–6 L/24 h IV antibiotics. Stage 3: sedation/anti-epileptics, vitamin K, FFP	Hunt et al. [11]
Kerrytown Sierra Leone LATE Rural	ETC	UK MoD	Referred positive cases. Health care workers only initially. Mortality 44%	High. Labs available. Monitoring-CVP line under ultrasound guidance (to reduce repeated venepuncture), catheter and bowel management	Clay et al. [12] O'shea et al. [13] Nicholson-

				system. IV fluids on all, I-O and vasopressors available. Blood products administered -FFP, cryoprecipitate, platelets, convalescent whole blood and packed red cells, tranexamic acid and vitamin K. Anti-helminths, anti-malarials and wide selection anti-microbials. Symptomatic management—analgesia, ranitidine, omeprazole	Roberts et al. [14]
US and Europe Whole outbreak	15 hospitals with isolation room in 9 countries	Local country institution	24 of 27 cases contracted in West Africa. 18 already started treatment while in WA 18.5% mortality	High. Treatment according to local protocol. Nearly all received IV fluids and, electrolye supplements. Nine received non-invasive and invasive vent, five renal replacement. Fifteen parenteral nutrition. Only 15% (4 pts) did not receive experimental interventions. Antibiotics, antimalarial, antiemetic, few anti-diarrhoeal. Also, blood, FFP, and platelet support. Of five critically ill with multiorgan failure—invasive vent and CRR, three died. Eight received vasopressors, five died. Trans cut pacing-1, none CPR	Uyeki et al. [15]

EHU Ebola Holding Unit, *ETC* Ebola Treatment Centre, *ECC* Ebola Community Centre, *WA* West Africa, *POC* point-of-care, *DFID* Department for International Development, *MoH* Ministry of Health, *WHO* World Health Organisation, *KSLP* King's Sierra Leone Partnership, *ORS* oral rehydration solution, *MSF* Médecins Sans Frontières, *RDT* rapid diagnostic test, *NSAIDs* non-steroidal anti-inflammatory drugs, *IV* intravenous, *SL*, *FFP* fresh frozen plasma, *CVP* central venous pressure, *I-O* intraosseous, *CPR* cardiopulmonary resuscitation, *vent* ventilation

patients who made it as far as hospital settings were included) most likely explain most of this variation, rather than the effect of differing management strategies. Therefore, drawing a meaningful conclusion from these mortality trends about the effect of treatment strategies is very difficult, though expanded access to staffing, therapeutic monitoring, blood product support, and critical care is almost certainly contributory to improved survival.

There was a range of the care delivered from differing facilities across the world. As aforementioned, the most recent WHO manual outlines categorizing the patients into mild, moderate and severe, however, this was not available at the start of the outbreak and many centres used their clinical judgment to determine appropriate therapy. Additionally, there was variation across laboratory testing, diagnostic abilities, clinical management and discharge criteria. This was due to multiple factors including the human resource and funding allocation, geography of the unit and specific patient factors (such as time of day they presented and agitation levels) which affected the safety of procedures.

A key component to the MOHS operational plan for country preparedness was the establishment of isolation facilities at every hospital admitting suspected, rather than just confirmed cases. EHU's such as at Connaught Hospital in Freetown provided a point of access to basic health care where unfortunately in practice the highest levels of management generally only reached intravenous lines and IV fluid and antibiotic therapy, due to resource constraints, leaving laboratory testing and closer monitoring to the ETCs. Isolation in these facilities was expected to be for less than 24 h, while waiting for blood results and transfer, however, patients were often waiting much longer due to delay in diagnostics and limited bed availability and thus received basic care for this period of time. Additionally, at the height of the outbreak when the number of suspected cases exceeded bed availability, there were many patients who were forced to wait outside hospitals, ETCs or at home, receiving only ORS, until a beds became available. It is clear now that this time window was critical for patient resuscitation and the ability to check laboratory results and correct electrolyte imbalances would have been of significant benefit.

Even once a patient reached an ETC, the level of care differed widely. For example Kenema ETC, in the East of Sierra Leone, one of the first ETCs set up by MSF and a 5 h drive away from the EHUs in Freetown was able to perform basic laboratory tests but often was so overburdened with cases that optimal management was challenging. However, in Kerrytown, Sierra Leone, where the British funded two ETC sites, one staffed by Save the Children and the other through the Ministry of Defense, patients could receive a higher level of care from onsite laboratory services measuring and correcting electrolyte imbalances and a lower patient to staff ratio allowing more time and a safer environment for monitoring and delivering intravenous fluid resuscitation. Unfortunately, epitomizing the global delay in rapid response to the outbreak, Kerrytown opened in late November 2015 providing a mere two months of care during the height of the outbreak. In addition, the UK MoD ETC at Kerrytown limited its patients through selecting only health care workers who had become infected, and with a high level of resources were able to do all supportive interventions apart from intubation and ventilation and renal replacement therapy.

From the review conducted by Ukeyi et al. [15] on the care of 27 patients with Ebola in the US and Europe, it is clear that in all centres regular monitoring and laboratory testing were available and management was tailored to results. Five patients received renal replacement therapy, nine received non-invasive and invasive ventilation and transcutaneous pacing was conducted on one patient. None of the patients in any of the settings received cardiopulmonary resuscitation as it was deemed that once a patient had lost cardiac output, resuscitation would be futile.

4.1.4 Management and Controversies Surrounding Specific Complications (Table 4.3)

4.1.4.1 Fluids and Electrolytes

At the start of the outbreak there was reluctance to provide intravenous fluid therapy mainly due to the risk to health care workers and challenges surrounding observing patients for bleeding complications, and therefore oral rehydration solutions were relied upon. However, over the course of the outbreak there developed an understanding of the mechanisms of death from Ebola through shock, with high lactate (values reported as ranging from 4 to 10 mm/L [20]), hypoxia from late multi-organ failure, renal failure and electrolye imbalances, in particular hypokalaemia from diarrhoea. This led to a consensus that the basic principles of resuscitative supportive care should be applied with aggressive management of intravascular volume depletion, the correction of electrolytes and prevention of complications associated with shock (Hunt et al). In one well-documented case managed in a German Intensive Care setting, diarrhoeal output exceeding 8 L in 24 h was reported, with intravascular collapse, and was managed aggressively with fluid repletion (7–13 L per day). Intravascular leak did occur (pleural effusions, pericardial effusion, ascites) but overall, the treatment was successful [21]. Certainly Dallatomasinas et al. [5] noted that 95% of deaths occurred within 10 days of admission, when supportive therapy is most likely to impact survival. There still is, however, uncertainty on how much, what type and how fast, fluid should be given promoting suggestions that further research is required in these areas, to optimize simple interventions, including resuscitation with a pre-mixed Ebola specific fluid [22]. There were very few complications related to fluid overload in the West African setting, and Perner et al. [22] commented on minimal capillary leak and less hypoxaemia/ARDS than in Western sepsis, suggesting that perhaps not enough fluids were given as opposed to too much. Unfortunately, the challenges faced from high patient to staff ratios, limited time inside the unit and concerns over bleeding risks through IV cannulation, (although clinically significant gastrointestinal bleeding was reported in less than 5% [9]) meant that patients were likely not receiving optimal medication.

Electrolyte disturbances are common in EVD, although electrolyte monitoring was not widely available in West African holding units and treatment centres. Case

Table 4.3 EVD complications and interventions

EVD complications		Interventions			
Complication	Notes	WHO standard	Optimal care	Controversies/unknowns	References
Respiratory problems					
Respiratory distress	Common	Consider treatable causes. Monitor saturations, provide O_2 if <90%, if available	Ventilatory support; consider fluid overload	Reluctance to offer invasive ventilation in higher resource settings is controversial as case reports support use.	[15, 16]
Circulatory problems					
Dehydration	Common, may be severe	ORS, IV fluids if severe, monitor input	Precise fluid balance monitoring	Role for CVP monitoring; Aggressive IV therapy may lead to fluid overload; Type of fluid most efficacious; Role of loperamide for severe diarrhea if no evidence of paralytic ileus	[17]
Shock	Multi-factorial, uncommonly due to haemorrhage	Reverse hypovolaemia, treat sepsis	Inotropic support if fluid resuscitation ineffective	Amiodarone trialed at one NGO Hospital	
Haemorrhage	Variable presentation, rarely the cause of shock or hypovolaemia	Consider transfusion of whole blood, containing clotting factors, vitamin K, blood transfusion, fresh frozen plasma, tranexamic acid	Monitor Hb, HCT and coagulation and provide targeted treatment	Role of Vitamin K and transexamic acid	
Anaemia	Not common	Assess clinically	Measure Hb Treat if severe	Role of transfusion, threshold to treat	

Renal problems				
Acute Kidney Injury/Renal failure	Common in severe disease; associated with mortality	Monitor urine output, avoid NSAIDs	Renal replacement therapy (RRT) may be required, optimize fluid balance, treat reversible causes e.g. hypovolaemia	Reluctance to offer RRT in higher resource settings is controversial as case reports support use.
Acidosis	Common	Correct hypovolaemia and treat sepsis	Monitor bicarbonate, lactate, consider organ dysfunction	
Gastrointestinal problems				
Hepatitis/Liver failure	Hepatitis is common and associated with poor outcomes	Avoid concomitent hepatotoxic drugs	Monitor LFTs	
Dyspepsia +/− Gastric ulceration	Common	PPI e.g. omeprazole, ranitidine, Magnesium trisilicate		Not evidence based
Vomiting and diarrhoea		Antiemetics e.g. Ondansetron If bloody diarrhoea—metronidazole	Test for other infective causes for which targeted treatment may be given	Role of routine broad spectrum antibiotics	[9]
Neurological syndromes				
Anxiety/ Agitation	Common	Reassurance, environmental modification, Haloperidol/ benzodiazepines if severe	Psychological support	
Convulsions	Commonly reported-consider hypoglycaemia, hyperpyrexia and electrolyte disturbance	Anticonvulsants, Dextrose, electrolyte correction	Anticonvulsants	
Encephalopathy	Common—end stage	Extra nursing care, consider sedation, consider thiamine deficiency	Consider CSF analysis		[18, 19]

(continued)

Table 4.3 (continued)

EVD complications		Interventions			
Complication	Notes	WHO standard	Optimal care	Controversies/unknowns	References
Pain	Common	Analgesia: 1st line Paracetamol, 2nd line tramadol, 3rd line morphine; avoid NSAIDS/aspirin	Analgesia	Concern about respiratory depression with morphine	
Electrolyte disturbances					
Hypokalaemia	Common	Potassium supplementation orally+/− IV, IV magnesium	Regular electrolyte monitoring and replacement	As both hyper and hypokalaemia are possible, electrolyte monitoring is increasingly considered essential	[11, 12]
Hyperkalaemia	May accompany renal failure	Insulin and dextrose if severe	Consider need for RRT	As above	
Hypoglycaemia	May be common in children	Point-of-Care glucose monitoring ORS, oral glucose, IV dextrose may be required	Regular monitoring, Dextrose supplementation of IV fluids if required, Nutritional assessment/ support		

ORS Oral rehydration salts, *IV* Intravenous, *CVP* Central venous pressure, *HCT* Hematocrit, *NSAIDS* Non steroid anti inflammatory drugs, *LFT* Liver function test, *PPI* Proton pump inhinbitors, *CSF* Central system fluid, *RRT* Renal replacement therapy

series have highlighted variable potassium requirements in EVD. In a series from Kerrytown, Sierra Leone, 33% patients had potassium abnormalities—with similar proportions having hyper- and hypokalaemia, highlighting the need for point of care monitoring to direct therapy [11]. Hyperkalaemia was associated with acute kidney injury and mortality. In the same series, hyponatraemia was also prevalent, in over 30% patients.

4.1.4.2 Intubation, Ventilation and Haemodialysis

There has been controversy over the appropriateness of intubating and ventilating patients with Ebola. In November 2014, a meeting was convened by the Chief Medical Officer of the UK, Dr. Sally Davies, to review the evidence surrounding critical care for patients with EVD. Subsequently, a statement was published concluding that there was "no evidence that addition of ventilatory or renal support would result in substantial overall benefit for patients who receive the optimum supportive care for Ebola" (Jacobs M, Beadsworth M et al. Provision of Care for Ebola. Lancet 2015, https://doi.org/10.1016/S0140-6736(14)62250-9. However, this was met with a rebuttal [16, 23] summarizing the evidence from case reports in USA and Germany, where five of eight patients who received critical care interventions made a complete recovery. Those who died had received critical care late in their illness. Additionally, the CDC and international nephrologists stated that performing haemodialysis in Ebola was safe, with no virus detected in dialysis effluent [24]. The UK's Ebola referral hospital, the Royal Free, is now prepared to perform haemodialysis and intubation. There are still no standardised guidelines for use in high income settings, however, the current recommendations are that organ support should be available but that interventions should be limited with a do not resuscitate (DNR) order place.

4.1.5 Specific Treatments

There are no specific treatments with proven efficacy for Ebola. The rapid and unpredictable onset of EVD outbreaks, together with their often remote location and hazards to health workers prove a major challenge to therapeutic trial implementation. A series of eight symptomatic patients in Kikwit, in 1995, given convalescent whole blood, was the largest reported attempt at a an Ebola therapeutic intervention study, prior to the West African Ebola outbreak [25].

However, this has been a rapidly evolving field. The scale and duration of the West African EVD outbreak, involving patients in urban and high resource settings, provided the opportunity and incentive for accelerated research into therapies and clinical trials.

Efficacy studies to date have consisted of:

1. In vitro and animal studies
2. Case reports/small case series of occasional or "compassionate" human use
3. Phase II clinical trials

Interventions have been either post-exposure prophylaxis (PEP) in asymptomatic exposed individuals, and/or treatment in symptomatic cases.

Candidate drugs that have made it into human clinical use currently fall into four categories:

1. Novel specific antiviral agents e.g. small interfering RNAs
2. Repurposed anti-infective agents, e.g. favipiravir
3. Other repurposed agents e.g. amiodarone
4. Specific immune therapies, e.g. convalescent plasma, convalescent whole blood, monoclonal antibodies such as ZMapp

In summary, an impressive effort to generate human trial data on therapeutic inventions, the results of which are summarized in Table 4.4 (taken from Brown [34]) has to date identified no clearly efficacious agent.

There was a reluctance to attempt double-blind, placebo-controlled trials, for questionable ethical and logistical reasons (see Lanini [35]—this includes a good review of clinical trials and a discussion of the ethical issues). An opportunity to collect data on the efficacy of supportive interventions e.g. threshold for IV fluids, loperamide for diarrhea, was largely missed due to inadequate human resources, and no systematic controlled studies of such were performed to our knowledge. As EVD is most likely to affect the low income setting in future, further research into interventions, such as the safety of intravenous lines, usual care versus laboratoryguided care, effectiveness of anti-diarrhoeals, empirical antibiotics and oral potassium supplements would be beneficial. Provisional plans for trials should be agreed prior to the next outbreak, to improve rapid implementation when the need arises.

This is a moving field and more trial data is likely to be available in the upcoming years. We suggest consulting the following useful resources for up-to-date information:

https://ebolaclinicaltrials.tghn.org
http://www.ukcds.org.uk/resources/ebola-research-database
http://www.who.int/medicines/ebola-treatment/emp_ebola_therapies/en/
https://clinicaltrials.gov

4.1.5.1 Novel Specific Antiviral Agents

Small interfering RNAs, have shown potential in treatment and prevention, in non-human primates and humans. TKM-100802 (by Tekmira), was fast-tracked due to encouraging animal and phase 1 data for a clinical study in Sierra Leone. However, the trial was terminated early, as demonstration of efficacy was deemed unlikely, and the product discontinued by the manufacturer [29]. Other siRNA molecules are in development with encouraging data efficacy in non-human primates (limiting symptoms, abrogating mortality—siEbola-3, see Thi [36]).

Brincidovir, a small molecule nucleotide analogue, orally available lipid conjugate of Cidofovir (Chimerix, Inc.), was used occasionally in repatriated patients and in an open-label, single arm phase II clinical trial (in Liberia) which stopped early,

Table 4.4 Summary of published evidence available for the use of novel therapeutic agents with potential anti-EVD activity

Trial (References)	Design	Sites	Patients	Enrolment dates	End point reached	Outcome
Brincindofovir [26]	Single-arm phase 2 trial	1	4	January 1, 2015 to January 31 2015	No—trial terminated	All four enrolled patients died
Favipiravir-JIKI [27]	Single-arm phase 1 trial	4	126 (540 historic controls)	December 17, 2014 to April 8, 2015	No—reported differently	Nuanced conclusions; limited tolerability
Favipiravir-Jui [28]	Single-arm phase 2 trial	1	39 (85 historic controls)	November 1, 2014 to November 10, 2014	Not applicable: reported as a retrospective clinical case series	Survival rate 56% [22/39] intervention versus 35% [30/85] controls; $P = 0.027$
TKM-Ebola [29]	Single-arm phase 2 trial	1	14 (3 cohorts, observational)	March 11, 2015 to June 15, 2015	Yes—stopped due to futility	No survival benefit
ZMapp (PREVAIL II) [30]	Randomised-controlled (non-blinded) trial	11	36 (35 controls); Guinean patients received favipiravir; unclear if matched	March 1, 2015 to November 1, 2015	No—stopped early due to low EVD case numbers	Mortality rate 37% [13/35] intervention versus 22% [8/36] controls; posterior probability = 91.2%
Convalescent plasma [31]	Non-random comparative study	1	99 (507 controls)	February 17, 2015 to August 3, 2015	No—also uncertain if neutralising antibody present	Mortality rate 38% [158/418] intervention versus 31% [26/84] controls; $P = 0.92$ after age/Cycle Threshold value (CT) adjustment
Convalescent whole blood [32]	Non-random comparative study	2	44 (25 controls who consented for the study but did not agree to the intervention)	December 2014 to April 2015	Not applicable: reported as a clinical case series	Unadjusted case fatality rate 27.9% intervention versus 44% control
Interferon β-1a [33]	Single-arm, proof-of-concept study	1	9 (21 historical controls and a further 17 historical controls matched for age and baseline viraemia)	March 26, 2015 to June 12, 2015	No—low EVD case numbers limited the trial to a pilot study	Unadjusted hazard ratio for treatment = 0.27 [95% CI 0.08–0.94]; $P = 0.039$. However, increased to 0.79 when adjusted for viral load.

ORS Oral rehydration salts, *IV* Intravenous, *CVP* Central venous pressure, *HCT* Hematocrit, *NSAIDS* Non steroid anti inflammatory drugs, *LFT* Liver function test, *PPI* Proton pump inhinbitors, *CSF* Central system fluid, *RRT* Renal replacement therapy

after allocation of only four patients, after termination of manufacturing by Chimerix [26].

4.1.5.2 Repurposed Drugs

Favapiravir (Toyama, Japan), is a broad spectrum anti-viral, approved for treatment of influenza, in Japan. A therapeutic effect was reported in Ebola virus-infected mice and supported by relatively good outcomes in some cases of occasional human use in repatriated patients (PEP and treatment). Since then two historically-controlled, open-label, single arm phase II clinical trials have been undertaken: a multi-centre study in Guinea (the JIKI trial, see Sissoko [27]) and a single-centre study in Freetown, Sierra Leone [28]. Conclusions are limited by the chosen study design and the challenges of trial implementation under the circumstances. The former, larger study did not demonstrate a treatment effect of favapiravir on mortality, whereas the latter reported survival benefit and viral load reduction in treated patients. Together these results support prioritization of a randomized placebo-controlled trial of Favapiravir, when the opportunity presents.

4.1.5.3 Specific Immune Therapies

There is some evidence (and a widely held belief) that an Ebola-specific antibody response is protective, providing the rationale for attempted treatment and prevention with Ebola virus-specific antibodies. Specific monoclonal antibodies were available but in short supply and were not easily amenable to rapid scale up. One such antibody cocktail, ZMapp, was initially used compassionately in repatriated patients and when increased supply became available, the only randomized control clinical trial was performed, in US, Guinea, Liberia and Sierra Leone. The results suggested that there was some evidence of treatment effect, and again this agent will certainly be a likely candidate for prioritization in any future outbreaks [30]. Other monoclonal antibodies, (ZMab and MIL77) were also used in selected cases as PEP [37]. Due to encouraging results during the earlier Kikwit outbreak (albeit in selected patients, see Mupapa [25]) there was considerable enthusiasm for a trial of convalescent plasma and/or whole blood from EVD survivors. Phase II trials were commenced in Guinea and Sierra Leone, but neither trial recruited enough patients, and the Guinea trial did not ascertain whether any neutralizing antibody was present [31, 32]. Additionally, a pilot study of interferon β-1a for EVD enrolled only 9 patients in the intervention arm (reference Konde et al Plos One 2017 Interferon β-1a for the treatment of Ebola virus disease: A historically controlled, singlearm proof-of-concept trial. It remains to be seen whether there are long term side effects or other consequences of therapeutic manipulation of the natural immune response to EVD.

4.1.6   Conclusion

It is difficult to assess the effect of different interventions on outcomes, due to the lack of formal studies involving rigorous prospective data collection and appropriate controls. The epidemiology of the disease and the response changed significantly, and so historical comparisons are relatively unhelpful. We have concluded that, in the absence of evidence and the lack of opportunity to generate evidence, it should be the priority to optimize supportive therapy, including intensive care therapy when available. Other specific interventions may be used when they are readily accessible, likely to provide benefit based on extrapolation from other diseases, and very unlikely to do harm, in pragmatic studies. Other interventions, including novel drug therapies and immunotherapies should be evaluated in randomized, adequately controlled clinical trials, prior to use.

4.2   Ebola Virus Disease in the Obstetric Population

Diana Garde and Emily Headrick

4.2.1   Introduction

Care for the pregnant and postpartum patient in the context of an Ebola Virus Disease (EVD) epidemic presents unique and challenging clinical, ethical and logistical considerations. Since the first recognized epidemic of EVD in Zaire in 1976, subsequent outbreaks have occurred in Sub-Saharan Africa (SSA) [38] where access to robust maternal health care services has been limited. An EVD epidemic superimposed on an already weakened health system further exacerbates extremely poor outcomes for the obstetric population. In general, 15% of all women are expected to experience obstetric complications [39]. The 2013–2015 West African EVD epidemic is the largest to date, with approximately 27,500 cases in three countries—Guinea, Sierra Leone and Liberia. Prior to the 2013–2015 EVD epidemic, these countries demonstrated some of the worst health care indicators due to inadequate infrastructure, human resources and lack of access to basic medical resources [40–43]. In Sierra Leone, it is estimated that prior to the EVD epidemic, one in every 21 women would die in childbirth-related incidents in her lifetime, [44, p. 34] with the maternal mortality ratio at 1630 per 100,000 live births in 2010—the highest in the world [45]. In addition, all three countries fell within the lowest density of medical personnel per capita, resulting in the smallest and least-skilled healthcare workforce in the world [46]. Both the number of in-hospital deliveries and cesarean sections declined as the incidence of EVD increased [47]. Ultimately, it was predicted at the end of the epidemic in West Africa, that the greatest impact of EVD would be on maternal health due to the loss of medical staff to Ebola (physicians, midwives and nurses). By September 2015, the three countries had lost 513 medical personnel from their already significantly small workforce [48]. From that data, it was estimated that there would be an increase of at least 38% (Guinea) and as much as 111% (Liberia) in maternal mortality [49].

It has been asserted that women are at an increased risk for EVD infection, both physiologically and socioculturally. Traditionally, they are the caregivers for ill family and community members, and they responsible for burying the dead, and as such have significant potential for interaction and contact with the virus via body fluids [50]. If a woman is pregnant during an EVD epidemic, routine health care encounters for antepartum or intrapartum care places women at high risk for exposure to infected patients also seeking care. There is a higher prevalence of female medical staff, namely nurses and midwives, both inside government facilities and in the community who would then be more likely to have (knowingly and unknowingly) direct contact with positive EVD patients in an epidemic. In addition, females in SSA have historically had less authority over their reproductive health and are at higher risk of gender based violence or coercion [51]; sexual intercourse with an infected or surviving male also places women at increased risk of infection [52]. Natural immunosuppression occurring in pregnancy [53] may play a role in a higher acuity of illness observed in presenting obstetric patients with positive EVD, contributing to very poor maternal outcomes. Considering the expectation of an excess of 1.3 million pregnancies per year [54] across Guinea, Sierra Leone and Liberia, a significant number of women who had direct contact with the virus were either pregnant or postpartum.

During the 2013–2015 West African EVD epidemic, pregnant women presenting for health care services—either for routine peripartum care or in the event of obstetric complications—often met case definition for EVD (see Table 4.5).Where proactive clinical management would have been appropriate in a non-epidemic

Table 4.5 Clinical and epidemiological factors in initial EVD case definition screening

Clinical presentation	Epidemiological factors
Considerations for screening for general populations	
Early: fever, profound weakness or malaise, headache, myalgia, arthralgia, conjunctivitis, nausea or anorexia, throat pain or difficulty swallowing, abdominal or epigastric pain, diarrhea (bloody or nonbloody)	Exposure/Contact: infected animals, bushmeat or fruit also fed on by bats, healthcare workers/ traditional healers also treating EVD, items soiled or touched by positive EVD patient, deceased EVD bodies
Late: confusion and irritability, hiccups, seizures, chest pain, diarrhea (watery or bloody), vomiting (with or without blood), skin rash, internal or external bleeding, shock, respiratory distress	Sexual intercourse with EVD-positive male or EVD survivor
Additional considerations for screening obstetric population	
Vaginal bleeding of unknown origin, spontaneous abortion, premature labor and/or rupture of membranes, preterm labor, antepartum and postpartum hemorrhage, intrauterine fetal demise, stillbirth, loss of consciousness	Exposure to products of conception or deceased fetus of EVD positive patient Being a pregnant woman with history of contact with confirmed EVD patient, recent EVD survivor with an intact pregnancy, newborn of an EVD positive mother, infant breastfed by a recent EVD positive mother

WHO Library Cataloguing-in-Publication Data [55]

setting, women were often left untreated or were provided minimal intervention by frightened medical staff working in an overwhelmed, under resourced health care system in crisis. Even as the emergency response resources became more established, an unfortunate number of women were isolated based on presenting symptoms; the majority of these women ultimately tested negative for EVD, yet they languished and died in EVD isolation centers due to the limitations of obstetric interventions available in this setting. *It is in this context that clinical care for obstetric patients in an EVD setting be specifically considered.* Culture, health infrastructure, access to resources, geography and physiology are all complex forces that must be considered when managing individuals and populations alike. Treatment of the pregnant or postpartum patient who meets case definition for EVD should be offered without haste and with the highest level of quality interventions possible, while maintaining safety of healthcare workers and the community.

This chapter aims to achieve the following objectives, informed by previous research on EVD outbreaks and by field experiences from the 2013 to 2015 West African EVD epidemic:

- Examine the unique risks and needs of the obstetric patient in the context of the an EVD epidemic
- Review clinical management of the obstetric patient infected with EVD as well as the unconfirmed or suspect obstetric patient being managed in a Red Zone (RZ) setting
- Provide technical guidance for establishing a safe and effective setting for triage, screening, isolation and treatment of the pregnant patient and mother/baby dyads.

4.2.2   Clinical Presentation of EVD in Pregnant Women

4.2.2.1 Signs and Symptoms

One of the key challenges when evaluating a pregnant patient in the context of EVD is that symptoms suggestive of EVD infection (i.e. abdominal pain, vaginal bleeding) may mask a true obstetric emergency. Conversely, the clinical presentation of what initially appears to be an obstetric complication often confounds the accurate identification of potential EVD infection. However, EVD infection can be the precursor to obstetric complications and a patient presenting for care may in fact require appropriate interventions for both.

The basic evaluation of suspected EVD patient must consider both clinical presentation (signs and symptoms) *and* an accurate history to evaluate risk of exposure. Due to this unique overlap in symptom presentation for the pregnant patient, additional considerations for the obstetric population must be made in addition to the general screening (Table 4.5).

It is of note that not all EVD-infected patients presented with or manifested a febrile state during their illness. It is estimated that only between 66% [56] and 89% [57] of patients actually presented with fever greater than 30 °C. Likewise, patients did not regularly demonstrate coagulopathies or overt hemorrhagic symptoms [57], as is frequently assumed as a common presentation for a viral hemorrhagic fever, with only one study finding these as a presenting symptom in only 35% of patients [56]. In the 2013–2015 West African EVD epidemic, it was frequently observed that EVD-positive pregnant patients were presenting with tachypnea and tachycardia, sometimes in the absence of any other symptoms, and could be repeatedly seen "hitting the floor," or favoring a position on the floor rather than in bed. It is postulated that this behavior may be due to agitation, or to alleviate fever by lying on a cool surface, but was routinely observed among this population [55]. Although many guidelines for screening include fever and bleeding as dominant symptoms in the algorithm, suspect cases can present without either and EVD illness can be missed during initial clinical evaluation. In fact, there are multiple documented cases of pregnant women with known EVD exposure and/or infection who do not manifest these common symptoms as expected.

In these cases, the potential for an EVD positive fetus/neonate is 100%, thus increasing the risk to unsuspecting healthcare workers during routine intrapartum care [58].

The majority of obstetric complications outside of an EVD epidemic involve the following: postpartum hemorrhage (PPH), antepartum hemorrhage, obstructed labor, postpartum sepsis, complications of abortion, severe preeclampsia or eclampsia, ectopic pregnancy and ruptured uterus [39]. In 2010, approximately 440 women in SSA died each day as a result of obstetric complications, the majority from PPH [59]. In addition, the stillbirth rate in SSA is approximately 10 times higher than in developed countries at 29/1000 [60]. As such, an EVD-negative obstetric patient can easily meet current WHO case definition, resulting in the isolation and testing of a high proportion of women who are actually in need of basic obstetric care.

Given that overlap of presenting symptoms may be commonly seen in both EVD and non-EVD obstetric cases, astute clinical decision making is required to ensure that obstetric patients are not being overwhelmingly screened and disproportionately isolated for EVD when the absolute risk in the community is low. Conversely, great effort must be made to ensure that symptomatic obstetric patients are not being treated outside of isolation when the absolute risk of EVD is high. In West Africa, it was estimated that only 1.5% of the pregnant patients admitted to isolation were EVD positive; the 98.5% who were not merely needed intervention for obstetrical complications or normal delivery [61].

4.2.2.2 Epidemic Screening Case Definition

Epidemic: In an epidemic setting it can be argued that the likelihood of infection transmission or the public health risk from EVD is greater than the risk of morbidity or mortality from obstetric complications. You must first consider general screening for the basic symptoms of EVD infection, with additional symptoms to be considered for the obstetric patient (Table 4.6):

Table 4.6 Additional symptom screening criteria for the obstetric population, active EVD epidemic

General screening	Obstetrical considerations

Does the patient have:
  Fever and 3 or more symptoms
  History of contact and fever
  History of contact and one or more
  symptoms
	Or, any of the following:
	  Spontaneous abortion (SAB)
	  Pre-labor rupture of membranes
	  Preterm labor
	  Ante, intra or postpartum hemorrhage
	  Intrauterine fetal demise (IUFD)
	  Stillbirth
	  Neonatal death
	  Survivor with intact pregnancy in labor or
	  postpartum
	  Pregnant and contact (asymptomatic)

WHO Library Cataloguing-in-Publication Data (2016) [55]

Any of these additional obstetrical complications, particularly in combination with symptoms included in routine EVD screening, should warrant isolation and polymerase chain reaction (PCR) testing. In addition, screening methods must capture any patient who presents to a healthcare facility who has recovered from EVD with an intact pregnancy and is in labor or in the immediate postpartum period. Given the absolute likelihood of viremia in products of conception from a patient who recovered from EVD while pregnant, delivery must take place in isolation.

4.2.2.3 Non-epidemic or Post-epidemic Case Definition

In a post-epidemic setting the general screening criteria changes, but the obstetric screening considerations remain largely the same. Contrary to epidemic settings, the risks of poor obstetrical outcomes are overwhelmingly thought to outweigh the risk of EVD infection or transmission of disease; strong clinical judgment should be employed.

In the general population, initial screening uses fever as a determinate for case definition. This tool is used for any acute hemorrhagic fever (Table 4.7).

Table 4.7 Additional symptom screening criteria for the obstetric population, non-epidemic/post-epidemic

Non-EVD epidemic general screening	Obstetrical considerations

Does the patient have:
Persistent fever for 2+ days despite
treatment?
  And unexplained bleeding, or
  Clinical suspicion
	Or, are there any of the following:
	  Spontaneous abortion (SAB)
	  Pre-labor rupture of membranes
	  Preterm labor
	  Ante, intra or postpartum hemorrhage
	  Intrauterine fetal demise (IUFD)
	  Stillbirth
	  Neonatal death
	  Recent survivor with intact pregnancy in labor or
	  postpartum
	  Pregnant and known contact with someone with
	  active EVD (even if patient is asymptomatic)

Clinical Guidelines/Wall Charts. Adapted from revised WHO and MSF guidelines Oct/Nov/Dec 2014/Jan/Sept–Oct 2015) Revised Oct 30, 2015

The fever must have persisted despite treatment with appropriate medications for the symptoms (i.e. treatment for presumed malaria), prescribed by a qualified medical professional. The patients meeting criteria must be isolated and PCR tested.

Pregnant women pose a challenging dilemma during the initial months after an epidemic has been declared over. Those who conceived during the epidemic could potentially carry an EVD infected fetus into a post epidemic setting given the average 40 weeks of gestation. In a post-epidemic setting however, one might argue for a higher level of scrutiny and clinical judgment prior to admission. The singular obstetric complication alone as criteria for admission for EVD testing cannot be justified given the numbers of patients that present for care for PPH, IUFD and SAB under non-epidemic conditions.

4.2.3 Establishing Triage and Screening

4.2.3.1 When to Isolate

In either an epidemic or post-epidemic setting, obstetric patients with *highly concerning* presenting symptoms should be admitted for testing in isolation, regardless of whether they fit the established criteria. Consultation and collaboration is recommended with other obstetric providers, surveillance officers, and EVD medical professionals who should be made aware of these cases.

4.2.3.2 General Principles of Triage and Screening

The evaluation and planning for a patient presenting with symptoms of EVD is wholly dependent on whether there is a declared epidemic present, or if there are definitive laboratory-confirmed positive cases in the population. Where there is any concern for EVD or other hemorrhagic fevers, triage and screening should be established at the entry point to every tier of health care facility in the affected area per guidance of government mandates or international recommendation.

4.2.3.3 Screening Upon Presentation to a Clinical Facility

During a declared epidemic or in cases where there is a high suspicion of infection, an obstetric-specific screening area should be staged at all peripheral health units, community clinics, and hospitals (i.e. anywhere obstetric patient may present). All pregnant, postpartum and lactating women should have an accurate temperature taken, and basic and obstetric-specific screening questions addressed before entry to the facility. Trained maternity nurses should complete the initial screening and evaluation to allow for experienced clinical judgment in this special population.

Ideally, screening should be operational for as many hours in a day as the facility is active. Staff should have clear protocols for documenting patient information to ensure multiple opportunities for confirmation of triage status upon entry into the health care encounter. We recommend standardized forms that can be made in duplicate or triplicate for record keeping. Staff should also have ways and means of communication to other colleagues so as not to leave the triage post in the event of a positive screen, an emergency in triage, or need for consultations or technical/operational support. This station should have access to electricity or reliable lighting, and a source of water for hand washing stations and cleaning.

While the triage area should ideally remain a calm environment for efficient yet thorough screening, the possibility of emergent screening is eminent. Triage staff should be prepared to prioritize emergent or critical cases while adhering to screening protocol and ultimately prioritizing safety for themselves, colleagues and patients in the immediate surrounding area. Patients may present to the facility in critical condition on foot, by private car or by ambulance from another facility. If the patient is unconscious or unable to give a clear history of the present condition, immediate isolation should be the most likely triage decision. If family members or an accompanying health care worker can provide a clear history that places the patient at a lower suspicion for meeting case definition, clinical judgement can be exercised judiciously.

A secondary screening area can be used when staff determines that a patient is going to isolation, but more history taking and assessment must be made. An open window with a one-meter buffer can be placed between the staff area and patient secondary screening room. A patient can be removed immediately from the primary screening area and obstetric-specific information can then be gathered from safely behind the window. If the situation warrants, nursing staff can use the space in secondary screening to allow for observation to assist in determining if the patient meets case definition. When a patient enters the secondary screening, it should be considered a Red Zone until suspect status is otherwise ruled out, or the patient is transferred into the appropriate isolation ward and the secondary screening cleaned.

In some situations, emergent or non-emergent, a thorough screening is not feasible prior to delivery and in cases of unknown history care must be administered as if the patient meets case definition. It is beneficial to design and stock the secondary screening facility to immediately turn into a safe Red Zone in the case that a patient exhibits an emergent illness (i.e. actively bleeding, loss of consciousness) or in cases of imminent delivery. If the patient cannot be transferred into the secondary screening area, triage nurses must have the ability to close the primary screening zone from pedestrian traffic, offering privacy to the patient, room for nurses to work and separation from the general public until the emergency is resolved.

Clinical and IPC staff must have ready access to all necessary materials in the triage/secondary screening area to safely and effective care for patients requiring immediate attention. We recommend the following:

- Nurses must have a donning and doffing area and the means to handle blood borne waste or body fluids.
- Delivery kits, fluid resuscitation supplies, comprehensive PPE and medications for PPH should be maintained in the nursing area and access to a hospital bed in secondary screening is optimal for patient care.

- There should be the means for transferring a non-ambulatory patient to either isolation or to the hospital after the appropriate screening is completed, or postpartum. This may include a wheelchair or a stretcher.

4.2.3.4 Screening Within a Clinical Facility

There are known cases of pregnant women presenting to a health care facility in labor who do not meet case definition at admission, but who then develop higher-acuity symptoms during/after labor and are subsequently test positive for EVD [62, 63]. It is crucial that health care administrators, staff and supporting partners collaborate to establish and implement protocols to increase the chances of early identification, rapid isolation, adequate treatment and effective IPC in the health care facility.

Ongoing assessments must take place for the in-patient population throughout their stay in the hospital. There must be q-shift assessments completed to determine if a patient is showing signs of becoming ill or demonstrating the classical obstetric warning signs. Medical personnel must also screen neonates post-delivery q-shift, during their entire stay and discharge instructions must offered to caregivers regarding warning signs after discharge. Given the potential for a scenario in which a patient has subclinical asymptomatic infection, but passes EVD to the fetus, clinicians must be watchful for and proactive in isolating and treating as soon as symptoms arise.

4.2.4 Safely Managing Obstetric Emergencies and Deliveries in Green Zone Areas

During an active EVD epidemic, every patient must be treated as potentially infected, no matter where they may suddenly require care. While care should never be withheld from a patient requiring attention, appropriate IPC measures must be available before care is delivered. Given the rapid progression of obstetric emergencies and precipitous labors, rapid response protocols should be established.

Any area can be rapidly turned into an ad-hoc Red Zone. All first responders should be comfortable with IPC principles and with the concept of Red Zone/Green Zone, which can theoretically be established with or without physical barriers.

Multiple sets of complete PPE should be available at multiple locations in and around health care facilities and in ambulances. *Care should not be rendered without appropriate PPE and IPC materials.* We recommend delivery kits also be easily accessed or assembled for imminent deliveries. Support staff should be available to secure a perimeter around an ad-hoc Red Zone to maintain crowd control, for delivery of additional materials and medications, to establish a line of communication between ad-hoc Red Zones and surrounding areas, and to initiate the next steps of transfer into a facility once immediate care is rendered.

Hygienists/IPC support staff should don full PPE and prepare to decontaminate the ad-hoc Red Zone once the patient is stabilized and transferred.

In the event of potential exposure or contamination to others in the community, every attempt to identify potential contacts is crucial for ongoing contact tracing and surveillance. While this can be difficult when an ad hoc Red Zone must be established in a large facility, in a crowd, or involving methods of transportation, it is crucial in the midst of an outbreak.

4.2.5 Establishing an Obstetric-Specific Red Zone

The 2013–2015 West African Ebola epidemic catapulted the research and development of innovative designs for the structures and materials used to combat this virulent disease. While the relative quality of infrastructure used as Ebola Treatment Centers ranged from sticks and tarps to military grade modular isolation units to modified existing structures to multimillion dollar BSL4 biocontainment facilities at prestigious academic medical facilities in the United States, the guiding principles of infection prevention and control remain the same across sites.

The guiding principles of an Ebola Treatment Center include:

- Limited entry and exit points, for both patients, staff, materials and corpses.
- Double barriers between Red Zone and external environment.
- Flow of movement from Green Zone, to suspect case areas, to confirmed case areas, to morgue, thus reducing the risk of transmitting virus from confirmed EVD positive patients to possibly uninfected patients.
- Multiple points for decontamination between patients
- Adequate distance/barriers between patients and between "suspect" and "confirmed" wards
- Clinical principles that prioritize safety of staff, minimizing risk of contact with infected fluids.
- Around-the-clock bedside care is very difficult to realize given staffing constraints and limits of length of time in PPE. Regularly scheduled "rounds" to administer patient care is more feasible [64].

These principles remain the same for an Ebola Treatment Center serving the obstetric population. However, in our experience, we recommend the following considerations to the design and utility when caring for pregnant or laboring women in a Red Zone.

- ETCs are generally separated into suspect vs. confirmed wards, and/or "wet" versus "dry," meaning patients with active vomiting or diarrhea should be separated from those who do not to reduce the risk of transmission of high-viremia fluids. We recommend that in an Obstetric Red Zone, every attempt should be made to arrange patient beds so that actively laboring or unstable patients occupy an additional "ward," ideally one that can be visually monitored at all times. In the OB Red Zone setting, intrapartum patients are considered highest-risk in terms of potential of EVD transmission (in known positive or unknown PCR status), but regardless of their EVD status, will require the most focused care until they are stabilized.

- It is likely that staff will not be able to provide 24 hour support in the Red Zone, but particularly when a patient is actively laboring, every attempt should be made to schedule teams of two to rotate through "Red Zone rounds" to assist with labor without interruption of bedside care. Deliveries can be precipitous, and forgoing constant bedside care in active labor increases the risk that simple complications (i.e. shoulder dystocia, PPH) result in death.
- In the event that staffing support does not allow for constant bedside care during an active labor, we highly recommend designing the OB Red Zone with multiple access points to visualize patients from the Green Zone, either through windows or utilizing camera or video recording equipment if available. Being able to visually assess the status of a patient to determine the best time to don PPE for bedside care can be crucial to improving outcomes.
- Every Red Zone should be equipped with the appropriate materials to care for the obstetric patient. There are several key items that will be necessary to have in abundance:
  - Menstrual pads and diapers (adult and infant)
  - Infant formula and feeding cups
  - Cotton sheets (or suitable substitute) to cover patient for warmth and privacy, also to anticipate multiple bed-linen changes, and several rags, towels, linens for cleaning large amounts of blood/amniotic fluid.
  - Bassinette which can serve as sleep area for neonate or set up as neonatal resuscitation surface post delivery (must have cleanable surface)
  - Suture materials, needle holder, blunt tipped scissors, speculum or retractor
  - Manual vacuum extractor or forceps
  - Individual blood pressure cuff and thermometer at each bedside, able to be sanitized.
  - Sharps container at each bedside
  - IV poles, both and short (short for patient who must be placed on the floor for safety)
  - Bell, intercom or other means of calling for assistance. This is particularly helpful for women in early stages of labor who may not require constant monitoring to alert staff that assistance is needed.
  - Wall clock to monitor contraction intervals, as well as time limits for staff in the Red Zone.
  - Scale for infant weights or for measuring maternal blood loss
  - Ready-made and easily accessible kits with necessary equipment to rapidly manage normal deliveries, postpartum hemorrhage, and eclampsia/seizures

4.2.6 Medication Formulary

The acuity and complexity of patients entering into the Red Zone is such that a standard medication protocol is warranted. Given the lack of immediate diagnostic capacity in many settings, it is recommended that all patients receive empiric treatment of antibiotics, antiprotozoal and antimalarial medications until bacterial, amoebic or malarial infections can be definitively ruled out, or the entire course completed. Even in the case of EVD, one cannot rule out co-infection with malaria or other common infectious disease. As such, EVD suspect and positive patients should be continued on all medications unless testing confirms absence of co-infection.

Clearly, it would be optimal to have an extensive medication formulary at the disposal of clinical staff in an epidemic, however the historical outbreaks occurring in low resource countries have forced makeshift pharmaceutical supply. The following are suggestions for coverage of the potential needs in isolation or treatment centers. Planning should include at minimum medications from the following categories. All medications listed are from the WHO Model List of Essential Medicines [65], unless otherwise noted. Clinical judgment must be made as to the relative benefits and risks involved and the acuity of the individual patient when choosing medications (Table 4.8).

4.2.7 Protocols

4.2.7.1 Admission
For purposes of simplicity and as a model for the ideal facility, the following will be addressed as if both suspect and confirmed cases are in isolated and treated within the same site. At admission, it must be determined where the patient should be placed within the unit based on their status. "Wet" suspect patients, or those with active bleeding, vomiting or diarrhea should be separated from "dry" suspect patients. All suspect women awaiting test results should be physically separated from probable confirmed patients. Laboring patients should be given privacy and placed inside intrapartum rooms for delivery, containment of body fluids and a higher level of care and observation. Infants should be with their mothers and not in a nursery. Bassinet sharing must be avoided.

It is critical for every patient who enters into an isolation or treatment center, that complete demographic and symptom history information be completed at the time of admission for care planning, data collection and tracking purposes. First name, last name, and birthdate should be verified and patients given a wrist name band including those three identifiers. Patient information can be entered onto a whiteboard or other central documentation record for clinical planning with the following information: bed number, name (first and last), age, birthdate, pregnancy status (pregnant and gestational age, postpartum or lactating), presenting symptoms, and date of onset of symptoms. This record can also include pregnancy outcomes that are updated in real-time (i.e. delivery date and time, gender of neonate, complications, etc.). It is advisable to have an admission book, or means of keeping patient status updated and relevant data logged. The goal is that all staff members can quickly assess and interpret the status of all patients in the ETC, as status can change rapidly.

Another whiteboard or central documentation record can also be used to track lab tests completed (date and time), when the next confirmatory tests are due and results.

Table 4.8 Recommended medication formulary for the EVD Red Zone

Medication	Considerations/Indications for Use
Name, class	BF = breastfeeding safety CI = contraindication
Anesthetic/Analgesic	
Local: Lidocaine, 2% w/epinephrine	Short procedures
Injectable: Ketamine	Palliative, short procedures
Inhalational: Nitrous Oxide	Analgesic, Palliative care
Morphine (opioid), pain relief	Severe pain, end of life
Haloperidol (Haldol), antipsychotic	Anxiety, combative behaviour BF: with caution
Tramadol[a] (opioid), pain relief	BF: acceptable
Naloxone (Narcan)	For opioid overdose
Antipyretic	
Paracetamol (Acetaminophen), antipyretic, pain relief	PO/IV
Anxiolytic	
Diazepam (Valium), benzodiazepine	BF: with caution
Allergy/Anaphylaxis	
Diphenhydramine	PO/IV Optimal for blood transfusion
Epinephrine	
Anticonvulsant	
Diazepam (Valium), benzodiazepine	BF: with caution
Magnesium Sulfate	IM, potentially IV with appropriate monitoring capacity Use per WHO recommendations Only for use in severe pre-eclampsia or eclampsia
Calcium Gluconate	IM: For Magnesium toxicity
Antibacterial	
Amoxicillin and Clavulanic Acid (Augmentin), Penicillin	BF: compatible
TMP/SMX, Sulfamethoxazole and Trimethoprim (Bactrim)	Bacterial meningitis, sepsis, Shigellosis… CI: near term gestation, BF: compatible if healthy and term, CI: neonate jaundice
Ampicillin, penicillin	Ampicillin and Gentamycin recommended for chorioamnionitis[b] BF: with caution
Cephalexin (Keflex), cephalosporin	BF: compatible
Ceftriaxone, cephalosporin	BF: compatible
Azithromycin, macrolide	Chlamydial infections, PID BF: with caution
Doxycycline, tetracycline	Chlamydial, Gonorrheal infections BF: acceptable short term
Ciprofloxacin, fluoroquinolone	

(continued)

Table 4.8 (continued)

Medication	Considerations/Indications for Use
Name, class	BF = breastfeeding safety CI = contraindication
	Typhoid BF: compatible
Gentamycin[e], aminoglycoside	Ampicillin and Gentamycin recommended for chorioamnionitis[b] Gentamycin and Clindamycin recommended for endometritis[c] BF: compatible
Clindamycin, lincosamide	CI: diarrhea, watery or bloody stool BF: acceptable, use alternate if possible
Metronidazole (Flagyl)[e], nitroimidazole	Septicemia, giardia, suspected anaerobic infection BF: with caution, if benefits outweigh the risks, use alternative if available
Silver Sulfadiazine, sulfonamide	2nd and 3rd degree burns CI: late term: kernicterus risk
Antimalarial[d]	
Clindamycin and Quinine	Recommended for first trimester, uncomplicated malaria treatment
Artemether and Lumefantrine (Coartem), or other Artemisinin-based combination therapy (ACT)	Recommended for second and third trimester uncomplicated malaria treatment, PO—must take recommended dose for full three-day course, missed doses result in ineffective treatment
Artesunate	Recommended for severe malaria in all trimesters: IV or IM (must have at least three doses 12 h apart, dosage based on weight. ACT is then necessary to complete treatment
Antifungal	
Fluconazole	PO Avoid in early pregnancy BF: acceptable
Clotrimazole	Topical, vaginal insert Topical acceptable for pregnancy and BF
Antiviral	
Acyclovir	Herpes simplex virus, Varicella BF: compatible
Corticosteroid	
Dexamethasone, Betamethasone	IM, Maternal Recommended for preterm fetal lung maturation (26–35 weeks gestation PTL)
Antihypertensive	
Methyldopa (Aldomet)	PO
Hydralazine (Apresoline)	IV
Nifedipine[a] (Adalat), calcium channel blocker	PO BF: no adverse effects noted

(continued)

Table 4.8 (continued)

Medication	Considerations/Indications for Use
Name, class	BF = breastfeeding safety CI = contraindication
Labetalol, beta blocker	PO or IV
Furosemide (Lasix), loop diuretic	PO or IV Recommended for preeclampsia/eclampsia (antepartem, intrapartum, postpartum) related pulmonary edema
Antiemetic	
Metroclopramide (Reglan), prokinetic	PO BF: acceptable Can be used as galactogogue CI: depression
Ondansatron (Zofran), serotonin 5-HT3 receptor antagonists	PO or IV BF: no adverse data
Antidiarrheal/Laxatives	
Loperimide (Immodium)	May be considered for symptom management in EVD in the absence of bacterial GI infection [17] BF: with caution
Zinc Sulfate	For acute diarrhea
Oral Rehydration Salts	PO, As needed, in large supply
Docusate Sodium, stool softener	PO
Senna, bulk forming/irritant	PO
Uterotonics	
Misoprostol	SL, intravaginally, per rectum For incomplete SAB or TAB, or PPH where oxytocin is not available
Oxytocin	IM or IV Labor induction or prevention/treatment of PPH
Ergometrine	PPH
Nifedipine (Adalat)	Can be used to delay progression of labor BF: no adverse effects noted
Neonatal medications	
Tetracycline 1% ophthalmic, Erythromycin 0.5% ophthalmic	Prophylactic
Ampicillin and Gentamycin	Neonatal septicemia
Caffeine Citrate	Apnea of prematurity
Chlorhexadine	Recommended for cord care
Vitamin K	IM, Prophylaxis for coagulopathy
Supplements/fluids	
Folic Acid	400 mcg daily—for prevention of neural tube defects [a]Can be delivered in combined prenatal vitamin if available

(continued)

Table 4.8 (continued)

Medication	Considerations/Indications for Use
Name, class	BF = breastfeeding safety CI = contraindication
Iron	120 mg elemental
Potassium Chloride	Electrolyte loss
Glucose	To reverse hypoglycemia
Water for Injection, Sodium Chloride	To reconstitute injectable/IV medications
Lactated Ringers, Normal Saline	IV fluids for resuscitation
Antiseptics	
Chlorhexadine	
Povidone Iodine	
Disinfectants	
Alcohol-based hand gel	
Chlorine for solution	

Notes

A loading dose for antibiotics should be considered when creating a protocol for the isolation or treatment center

Decisions should be made regarding the protocol for empirical treatment based on available medications and specifics of local infection prevalence

NSAIDs and aspirin should be avoided in EVD settings due to risk of coagulopathies; NSAIDS also should be avoided in third trimester of pregnancy

No recommendations are given for Tuberculosis or HIV given the longer-term considerations and need for referral. No recommendations are given for anti-helminthic treatment given the relative low acuity of those infections

[a]Not in the WHO Model List of Essential Medicines

[b]Ampicillin and Gentamycin (chorioamnioitis) Hopkins 2002 (Cochrane Review), French 2004 (Cochrane Review)

[c]Gentamycin and Clindamycin (endometritis) French 2004 (Cochrane Review), Livingston 2003, Sunyecz 1998, Mitra 1997, Del Priore 1996, Barza 1996

[d][66]

[e]http://www.acog.org/Resources-And-Publications/Committee-Opinions/Committee-on-Obstetric-Practice/Sulfonamides-Nitrofurantoin-and-Risk-of-Birth-Defects 2011

Plans can then be documented (discharge home, transfer etc.) and reviewed or changed based on incoming results.

4.2.7.2 Assessment

In order to create a care plan for individual patients, a complete head to toe assessment must be completed to the best ability of the clinician, given time and patient load constraints. Information that should be collected can be divided into two categories—objective and subjective. A clinician with experience in obstetric care should collect the objective information. The subjective information is dependent on patient consciousness, the ability of a patient to be an accurate historian or in some cases must be pieced together from family members. The clinician inside the Red

Table 4.9 Clinical assessment of the obstetric patient

Subjective	Objective
Significant health/OB history Date of onset of symptoms History of present illness Medications/herbs taken prior to admission	Vital signs Level of consciousness Edema Bleeding Hydration

Pregnant
  GTPAL (gravida, term, preterm,
  abortion, living) status
  Gestational age/LNMP
  Fetal movement
	Pregnant
	  Fundal height
	  Fetal heart tones (if Doppler available)
	  Fetal movement (if over 20 weeks)
	  Ultrasound confirmation of viability, gestation, lie
	  (if available)

Postpartum
  Breastfeeding
  History of delivery (date, infant
  status)
  Significant delivery details
  including complications·
  Location of infant if viable
  If infant is living, status of infant
  (healthy, ill)
	Intrapartum
	Stage of labor (number of contractions, cervical dilation,
	effacement and descent for baseline reference)
	Membranes intact or ruptured, note quality of amniotic
	fluid
	Fetal status/fetal heart tones
	Moulding
	Medications administered in intrapartum period
	Immediate postpartum:
	Uterine tone
	Vaginal bleeding, source of bleeding
	Mental status

Zone is responsible for collecting gaps in history that has not been addressed before admission (Table 4.9).

All information must be documented in patient charts after each assessment in the Red Zone and on the central documentation record/whiteboard. Patients in labor should have a partograph started if over 4–6 cm dilation.

4.2.7.3 Testing Recommendations and Protocols

To date, PCR has been the testing method of choice for EVD infections. Approval for use of a Rapid Screening Test (RST) or GenExpert for EVD would allow access to results that could rule out EVD quickly. Choice of testing method must be made according to current international standards, national regulations and pharmacy board approvals.

All patients entering into the center should be PCR tested on admission. If the symptom onset is less than 72 h, then a second test is needed at or after the 72–h mark. Suspect and confirmed patients having either a spontaneous abortion or IUFD must have products of conception swabbed and tested (fetus, placenta—fetal side, associated tissue or amniotic fluid). An neonate delivered in isolation, or admitted with a suspect patient must also be tested.

A patient who is deceased prior to serum collection should have oral swabs collected prior to burial. IUFD at full term can also have oral swabs collected.

4.2.7.4 General Patient Care

Like any other EVD suspect or confirmed patient, pregnant, postpartum or lactating women must be assessed and treated based on presenting symptoms. Given the nature of fluid loss, offer appropriate replacement fluids (ORS or IV/IO) based on the severity of dehydration and level of consciousness. Consider differential diagnoses and offer antibiotic and antimalarial therapy as discussed above. Symptomatic relief must also be included in the care plan for pain, nausea, vomiting, agitation etc.

4.2.7.5 Intrapartum Care

Obstetrical care in the context of EVD has historically been limited to expected management given extremely high viral load present in amniotic fluid, blood, and placental tissue. As a result the inherent risk to health care workers it was determined to be too high for interventional care. The following were considered high risk in the past and healthcare workers were cautioned against engaging in: cesarean section, artificial rupture of membranes (AROM), episiotomy or deinfibulation of scarring related to female genital mutilation, manual vacuum aspiration, manual removal of a retained placenta, suturing, vacuum extraction, and craniotomy in the case of obstructed labor. Anticipated delivery in isolation should be managed with caution ensuring the utmost safety of staff while offering the highest level of care to the patient and her fetus. Adequate staffing numbers and skill level and immediate availability of needed delivery supplies and medications will eliminate some of the risks associated with deliveries in a limited resource setting. In serious cases where life is threatened or suffering is unmanaged, a higher level of intervention should be considered when qualified staff are present, adequate supplies are on hand and IPC measures can be adhered to. In this way, risks can be mediated and the potential for survival increased.

4.2.7.6 General Delivery Guidelines in the Red Zone

For management of all deliveries in the Red Zone, protocols must be created with respect to the local Ministry of Health clinical guidelines and WHO recommendations for OB care in limited resource settings. Additionally, the following considerations must be made to care for patients with EVD filovirus or other hemorrhagic fevers:

- An adequate intrapartum setting must be prepared before delivery to decrease the risk to staff and patients in the Red Zone. Safety must always be a priority and clinicians must not place themselves at risk in the event a lack of appropriate water, lighting, or PPE should occur.
- Red Zone staff should prepare the bedside when there is an impending delivery and have IV fluids, delivery kit, resuscitation equipment and neonate blanket and bedding at the ready.
- Staffing should be adequate for deliveries; ideally, there should be a nurse or midwife for the delivery, a nurse for the neonate and one other clinical staff member to monitor IPC and assist where needed. Presence of a hygienist is also advisable. Roles should be assigned prior to entry into the Red Zone.

- A second team in the Green Zone must be ready and able to relieve Red Zone staff when they are exiting and there must be a system of report to update status of the patient before the change of shift. If able, the exiting IPC clinician should update the incoming team before they enter.
- There must *always* be staff in the Green Zone to hand in needed medications, consult and support Red Zone clinicians. The Green Zone staff must help to monitor total time in the Red Zone and give adequate warning when doffing is required. A "sign in" whiteboard at the entrance to the Red Zone area allows for accurate monitoring of time in PPE and identification, location and duties of staff in the Red Zone.
  - It is optimal for clinical staff to visualize laboring patients from the Green Zone if there are not enough staff for round the clock care in PPE. A communication tool is also advised so that the patient can call for help if needed, or give status reports to staff.
- Elbow length gynecologic gloves are preferred for deliveries. In addition to standard PPE, a heavier and thick reusable apron is recommended for the delivering clinician to protect the front of the coverall from body fluid, and to reduce the movement of a lighter, thinner apron. Have all needed PPE in the Red Zone and readily available.
  - Treat all body fluids as potentially infected
  - Place IV before delivery if possible in anticipation of likely IVF resuscitation and to reduce risks to staff associated with an urgent/emergent IV placement.
- Fresh 0.5% chlorine must be available for immediate decontamination of soiled gloves and gown. Hands must be washed or outer gloves changes between procedures.
- Limit the number of vaginal exams during labor to the initial assessment and intermittent progress checks q 4 h if needed, or the to fewest number possible.
- A partograph and/or detailed charting and adequate care planning must be maintained so as to monitor labor progression and anticipate potential interventions for complications.
- Regularly monitor fetal heart rate or movement.
- Refrain from using fundal pressure during second stage
- A clear plastic sheet should be used as a drape during delivery to separate the clinician from the neonate and placenta. The delivering clinician is avoid sitting or standing at the end of the bed or between the legs of the laboring patient to limit contact with blood or amniotic fluid exposure.
- Limit the number of sharps in the Red Zone. Use blunt tipped scissors if available for cord cutting. Prioritize single use instruments over multi-use. If multi-use instruments are used, a system for cleaning must be employed. A rinse in 0.5% chlorine will ensure adequate decontamination. A fresh water rinse and immediate drying will delay corrosion of metal instruments. If available, an autoclave is optimal.
- Suturing should be available and employed only when adequate lighting and experienced staffing are available and only in cases where the patient is cooperative (not agitated) and there is a risk of negative outcome without intervention.
  - Blood loss and uterine tone should be monitored closely after every delivery, regardless of gestation.

- PO medications are preferred, but if needed progress to IV, IM or IO. All postpartum patients should have 10 IU of slow IV push or IM oxytocin (after confirmation of single fetus) to decrease the chances of PPH.
- If delivery of live baby (or with retained placenta), tie off and cut the cord under plastic. In cases of fetal demise, leave the cord intact, deliver the placenta and place both placenta and fetus together into a body bag, using recommended IPC measures.
- A team of postpartum clinicians should be available to monitor vitals and infant transition in the hours after delivery
  - Referral for HIV or TB testing or PMTCT services should be offered at discharge if there is a known or suspected secondary infection.

4.2.7.7 OB Complications in the Red Zone

Given the nature of a viral hemorrhagic fever superimposed on pregnancy, the likelihood of obstetric complications and coagulopathy is very high. Safety of staff members must be prioritized, however, we recommend the consideration that more interventions may be appropriately rendered to improve patient outcomes than were recommended in the 2013–2015 EVD outbreak.

- Adequate IV fluids must be on hand for volume resuscitation in high acuity cases.
- Consider induction of labor in emergent cases only; otherwise, defer induction if a suspect case is highly suspicious of having EVD. This is in an effort to reduce unnecessary exposure of staff to body fluids.
- Deinfibulation of type 3 female genital mutilation (FGM) is discouraged, but may be performed if it is considered a life saving measure. If proceeding with procedure, perform at the patient's side, under plastic sheeting and with adequate anesthesia or pain relief.
- Perform fundal massage (with support of lower uterine segment) and 40 IU of oxytocin in 1 L of IV fluids (60 gtt per minute) for initial PPH management and be prepared to offer higher levels of intervention if bleeding persists, including: external aortic compression and uterine balloon tamponade.
  - In cases of antepartum or postpartum hemorrhage, consider having an established system for basic blood typing within the unit and a process in place for collection and administration of blood products. No suspect or confirmed patient blood can leave the unit unless a process is in place to transfer to a EVD specific lab. Staff must be trained to type patient blood and to monitor for transfusion reactions. Inadequate training would disqualify a treatment center from transfusing.
- Patients who are suspect and acutely ill are often placed in situations that are life threatening while they wait for PCR results. Consider establishing a site for isolation surgery (dependent on trained personnel, equipment and adherence to IPC measures) so that patient needs can be addressed in emergencies and the option of cesarean section be made available.
- Consider administration of Nifedipine over Terbutaline (may have more serious side effects and requires higher level of monitoring) for delay of early onset preterm labor or onset of uterine hyperstimulation. Consider antepartum

corticosteroid (ACS) administration in cases where preterm delivery (26–35 weeks gestation) appears to be unavoidable.

- Proactively treat PROM or PPROM with prophylactic antibiotics.
- Do not delay treatment of hypertensive emergencies. Offer antihypertensives (PO or IV) if systolic exceeds 160 and diastolic exceeds 110 and monitor regularly. Address severe preeclampsia or eclampsia immediately with IM magnesium sulfate and continue until 24 h post-delivery or last seizure (whichever is later). The risk of fluid overload must be balanced with the propensity of EVD patients to be fluid depleted and anuric. Monitor for symptoms of pulmonary edema and defer diuretics unless severe edema [67]. Rapid delivery can resolve symptoms of eclampsia and induction or augmentation should be considered.
- In cases of IUFD, deliver as with a live infant under plastic sheeting. In many cases, symptoms of infection are present and mimic EVD, particularly if labor has been delayed or obstructed. Treat with antibiotics for chorioamnionitis and monitor. Induce or augment labor to expedite delivery if the risk of puerperal sepsis is high. Given the extremely high risk of fetal death with positive EVD mothers, pregnancies must be monitored and ultimately delivered in isolation or in an ETU. Even if the patient is recovered, her products of conception will have high viral load and must be treated as infectious waste.
- Spontaneous abortions must also be treated with suspicion and POC from suspect or positive patients treated as infectious waste.
- Vacuum delivery and episiotomy can be considered for obstructed labor when it is believed to be a life saving measure. IPC measures should be maintained at all times. Referral for cesarean should be made when available, if failed assisted delivery.
- In the case surgical intervention is needed and not available, all attempts at supportive care and potentially palliative care must be administered until lab results are returned. Transfer of a non-infected patient with confirmed negative PCR to non-isolation facility can be arranged when lab results are returned. This patient who has been identified as not having filovirus must be advocated for outside of isolation or treatment units and reassurance given to medical staff that they are uninfected and safe to be operated on.
- Confirmed EVD-positive patients with no access to surgery must stay in the treatment center until recovery and two confirmed negative results have been received. The fetus, regardless of gestation, will very likely not survive and will need to be delivered within the isolation facility.
  - Depending on the location of the epidemic, there may be access to high intervention medical care and drug therapy for neonates born from an EVD positive mother and though the fatality rate has historically been almost 100%, priority for should be given to advances in clinical care and vaccines. However, in the absence of adequate interventions, the option of therapeutic medical abortion before discharge should be offered if an EVD positive patient has an intact pregnancy, the patient lives far away from a EVD treatment center and there is a risk that she will not return for delivery (thereby risking further infection in the community).

4.2.7.8 Preparing for Care of the Neonate in the Red Zone

Given that the majority of women entering into the ETC will be EVD negative, the majority of care should mimic the expectations for infant care in a non-epidemic setting. Fetal monitoring during labor and delivery should be provided for early detection of complications and immediate care of the infant is essential to ensure optimal outcomes and reduce the risk of infant mortality.

- Immediately after delivery, vigorously stimulate and dry the neonate
- Oropharyngeal/nasopharyngeal suction only if secretions suction only if secretions are obstructing the airway; there is no need to suction a vigorous neonate, even if meconium present
- Perform basic assessment (airway, breathing, circulation) and offer resuscitation if needed
- Allow delayed cord clamping and cutting in non-emergent settings, tie and cut cord under plastic sheet to minimize contact with blood
- Clean and dry infant and allow skin to skin/kangaroo care and bonding if the mother is capable
- Assess initial APGAR score and then complete routine vital signs: respiration rate, heart rate (umbilicus or brachial pulse) and temperature q 30 min × 2 after delivery and q shift (at least with every assessment of mother thereafter). Complete newborn assessments as thoroughly as possible given time restraints in PPE and check for jaundice, tone, retractions and feeding. Stethoscopes cannot be utilized in a Red Zone.
- Ensure that the infant bassinet is covered with mosquito netting to decrease the risk of malaria infection
- Offer Vitamin K injection IM (1 mg) at birth, ophthalmic Tetracycline ointment bilaterally, and chlorhexidine cord care per WHO recommendations
- Delay washing for 24 h is possible and dress infant in weather-suitable layers (1–2 layers more than adults) to stabilize body temperature.
- Offer other treatments based on individual symptoms—consider IM antibiotic treatment for bacteremia for 3 days if concerns regarding infection

4.2.7.9 Care of the Neonate When Mother Is EVD Positive

Survival of neonates delivered to EVD positive patients in or out of an ETU is close to 100% mortality rate. At the end of the 2013–2015 West African epidemic, one documented infant of a deceased EVD positive patient survived after an intensive antibody, broad spectrum antiviral and antibiotic regimen [68]. One can assume that a neonate is exposed to filovirus in utero, during delivery or during breastfeeding (though it has appeared based on documented cases of live births that neonates seroconvert shortly after delivery with unknown data around the impact of breastfeeding [69]). Recommendations then include keeping the mother and infant as a treatment dyad and allow the mother to care for the infant if able. Communication with the mother about her EVD status and expected outcome of the neonate should occur as early as possible, with adequate psychosocial support as needed.

4.2.8 Breastfeeding

There are several factors that must be considered when reviewing recommendations for breastfeeding a neonate after delivery in an isolation setting and before maternal PCR results are obtained. In resource low areas, the risk of unsanitary water to make powder formula, the unavailability of ready to use infant formula (RUIF), the lack of hygienic means to sterilize bottles, or the prohibitive cost to families to obtain artificial feeding or animal milk products for the neonate for up to 2 years after delivery must be weighed with the immediate risk of continued breastfeeding and exposure to EVD. WHO recommends that in cases where the mother is symptomatic and awaiting results that her breastfeeding be suspended. In these cases a discussion with the patient about the risks, benefits and options must be had. Where the mother is unable to give informed consent one way or the other, family members must be brought into the conversation and clinical decisions be made also taking into account the current prevalence of infection in the community.

If the patient's result is positive, it can be assumed that the neonate will be positive. There are questions about the initiation of breastfeeding and whether the infant is already infected and would (1) benefit from any maternal antibodies, (2) will get an increased viral load through breastfeeding, or (3) be likely to die regardless of feeding. WHO recommends suspending breastfeeding and starting RUIF (ready to use infant formula) until breast milk samples test negative $\times 2$ by PCR testing after which the mother should be encouraged to initiate or resume breastfeeding exclusively for at least 6 months.

For patients who suspend breastfeeding until results are returned, interim supply can be maintained with pumping. However, all expressed milk must be treated as contaminated material and discarded per appropriate IPC protocols.

4.2.9 Neonate Considerations with Deceased EVD-Positive Mother

If the patient expires during or after delivery but prior to maternal serum testing, oral swabs of the corpse are to collected prior to burial. Collect a serum sample for PCR testing of the neonate as soon as possible , however note that the infant must be in the care of isolation facility for 21 days following delivery. A neonate can be infectious but present as being asymptomatic 3 days prior to becoming ill and even then symptoms are atypical or non-specific. Accounts of live newborns delivered to positive mothers were all documented to be deceased within the 19th day of life [69]. The 21-day cut off allows for monitoring for the entire contact exposure period.

If neonate is positive, treat per pediatric EVD protocol or arrange transfer to an appropriate treatment facility via local government regulations.

4.2.9.1 Psychosocial Support, Patient Education and Discharge Planning

The psychosocial impact of the 2013–2015 West Africa Ebola epidemic is an immeasurable burden for thousands of people. A common sentiment often expressed both from patients admitted to ETUs and health care workers alike was the dehumanization and trauma of being isolated. The human element of direct patient care was covered by multiple layers of plastic, only eyes visible behind foggy masks. While novel efforts to humanize health care workers rapidly spread by word of mouth among first responders, the experience of being cared for in an isolation center is undeniably traumatic. We strongly advocate for early and robust psychosocial resources as part of a comprehensive EVD response.

Care for the pregnant patient also requires unique psychosocial considerations. While practices may vary across cultures, childbirth is a global phenomenon laden with tradition, ritual and social norms. Universally, the laboring woman is at her strongest, and yet at her most vulnerable. Effort should be taken to understand and support the unique sociocultural norms surrounding pregnancy and birth in a way that maintains dignity in delivery while adhering to all safety and IPC protocol in the context of EVD care.

4.2.9.2 Discharge Planning and Patient Education

In times of crisis, it may be common for the quality of patient care to suffer as safety or allocation of resources is prioritized. We assert, however, that quality can be incorporated into emergency response service delivery in a way that does not jeopardize safety or waste valuable resources. Discharge planning and patient education are components of emergency response service delivery that may be neglected in the height of an emergency, but in the context of EVD, these practices can be a crucial component to breaking the chain of transmission as well as improving the overall quality of patient care beyond the ETU. The obstetric population has unique needs upon discharge from an EVD treatment or isolation facility that should be incorporated into the development and implementation of any program working with women of reproductive age.

Resources and protocols should be developed early so that staff are prepared to guide patients when they are discharged from the facility. All patients discharged from a Red Zone should receive clear instructions that they are, by default, considered possible contacts and should monitor their signs and symptoms for 21 days after discharge. Examples of unique obstetric patient pathways are summarized in Table 4.10.

Table 4.10 Discharge planning and education for the pregnant/postpartum patient

Patient	Discharge planning	Discharge teaching	Wrap-around services
Patient is EVD negative still pregnant upon discharge, with no complications	Where is the patient going after discharge? How will she get there? Is there a risk for EVD contact in their household? Where is the patient planning to receive ongoing prenatal care/ planning to deliver?	Signs and symptoms of labor, obstetric emergencies and EVD Where to present if these signs/symptoms occur in the future Recommendations for ongoing prenatal care	Psychosocial support Prenatal care/ delivery planning Nutritional support
Patient is still pregnant, EVD negative with complications (i.e. active labor, pre/eclampsia, obstructed labor, etc.)	Clinical judgement should be exercised about the safest place for the patient to deliver. If delivery is imminent at time of discharge, consider delivery if transfer of care to a tertiary facility cannot be arranged or deemed too risky Coordination with closest secondary or tertiary health care facility is paramount in the event of complications. Notify nearest facility in the event of imminent delivery or Cesarean section	The laboring patient should be consulted about the plan of care, and kept informed of changing condition. If possible, family members should also receive updated, and be available for discharge teaching while patient is in critical condition	Coordination with secondary/ tertiary level facilities Blood supply Ambulance transportation Materials for delivery (i.e. PPE, delivery kit, medication)
EVD negative patient is no longer pregnant upon discharge, or fetus not viable (i.e. SAB, IUFD)	Does the patient require close medical attention in a hospital? If so, plan for transfer of care to nearest facility Protocol for contacting patient and local surveillance authorities pending PCR results of fetus or POCs Assessment and appropriate management of patient's grief reaction at loss of pregnancy	Signs and symptoms of post-partum or post-SAB complications (i.e. puepueral infection), signs and symptoms of EVD infection[a]	Psychosocial support Transportation Access to family planning services if desired Access to health care if complications arise after discharge home

(continued)

Table 4.10 (continued)

Patient	Discharge planning	Discharge teaching	Wrap-around services
Patient is postpartum, EVD negative with infant	Does the mother and/or neonate require close medical attention in a hospital? Is the neonate stable (i.e. feeding, voiding, etc.)? Where will the patient/baby go after discharge, how will they get there? What is the family support situation? Is the home a safe environment for the dyad?	Extensive teaching should be dedicated to mother/child dyads after discharge, including newborn care, breastfeeding, vaccines, warning signs for complications both mother and baby	Coordination with newborn care provider should be arranged, if possible. Family planning Nutritional support Close follow up to ensure dyad has connected with local MCH resources
Patient is deceased, EVD negative infant is alive	Is the neonate stable or requiring close medical attention? Consider transfer to tertiary level facility where neonate can be monitored in hospital nursery. Is next of kin available and willing to assume responsibility of infant?	Extensive teaching for family members, if available, regarding newborn care, including signs and symptoms of complications, EVD Connection with local resources for ongoing support, if available	Psychosocial support for family of deceased patient Nutritional support Notify relevant authorities for child protection services for close follow up
Patient is EVD positive discharged when PCR is negative × 2	See next section, "Care for the EVD Survivor Upon Discharge and In Future Pregnancies."		

[a]Patient should be aware that these are similar presentations and may result in repeat isolation

4.2.10 Considerations for Safe and Compassionate Management for EVD Survivors with Subsequent Pregnancies

The development of clinical guidelines and practical health care delivery strategies to address the unique needs of EVD survivors presented grand challenges for the community serving Guinea, Sierra Leone and Liberia in the recovery phase of the 2013–2015 West African EVD epidemic. Even verifying the number of EVD survivors in West Africa is a daunting, occasionally political task, but estimates are currently that out of 27,500 documented cases of EVD, there are approximately 13,000 survivors [48].

To date, there are no strong data to approximate how many of those survivors have become pregnant since recovering from EVD, but it may be asserted that the pregnant EVD survivor represents the nexus of vulnerability. In 2010, *before the*

EVD epidemic, Sierra Leone had a baseline maternal mortality ratio of 1630 deaths out of every 100,000 deaths [45]; that number has almost certainly risen, with some estimates by as much as 30% simply by virtue of the decimated health care system and decreased utilization of services [70]. Combining these baseline maternal health indicators with the pervasive stigmatization and fear of pregnant EVD survivors lends an extraordinarily high risk for neglect and mistreatment of the pregnant EVD patient, despite the fact that there have been no data to show active virus in the amniotic fluid or products of conception in the subsequent pregnancies of women who survived EVD. It should be noted that the risk for stigma affecting care for EVD survivors is not limited to the West African nations where the outbreak occurred; a case study detailing management of a pregnant EVD survivor planning to deliver in the United States reports marked discomfort and concern from hospital staff despite no evidence of risk for transmission of virus [71].

Furthermore, there are limited data about subsequent pregnancy outcomes for women who became pregnant after surviving EVD, with a small cohort study in Liberia showing a slightly higher incidence of miscarriage or stillbirth in Liberian EVD survivors as compared to the overall rate in both the developed and developing world [72]. However, national data for baseline miscarriage/stillbirth rates are not available in Liberia, Sierra Leone or Guinea, making data specifically reflecting the EVD survivor population murky.

Recommendations addressing breastfeeding for EVD survivors in subsequent pregnancies were initially limited due to lack of data regarding viral persistence in breastmilk. Anecdotal evidence suggests that Ebola virus may persist for several months in breastmilk of survivors, but breastmilk was not routinely included in the major viral persistence studies conducted in the immediate post-epidemic period. There are no known cases of a breastfeeding infant presenting with EVD contracted from a lactating mother. Given the overwhelming benefit to breastfeeding, particularly in resource-poor settings, current CDC guidelines support routine breastfeeding of the neonate born to EVD survivors, with case-by-case evaluation to neonates born to suspect or confirmed EVD patients [73].

There are multiple circumstances complicating the approach to and delivery of quality care for the pregnant EVD survivor, largely due to emerging research regarding viral persistence, clinical sequelae in EVD survivors and sociologic trends of stigma and access to resources for pregnant EVD survivors in their communities.

With the data available, we support the following recommendations:

- EVD survivors presenting with subsequent pregnancy outside of a known EVD epidemic should be treated as a non-infected patient. Antepartum and intrapartum care should not be delivered with any more PPE than would be used for a non-survivor patient (universal precautions).
- Status as an EVD survivor should be considered as a relevant part the patient's history to inform any abnormal clinical presentation, and treated with astute clinical judgement. Other key components of a thorough history taking for a pregnant EVD survivor include: survivor status of the father of the baby, any recent illness/complications, social support status.

- The heightened vulnerability of a pregnant EVD survivor should inform a broader and more comprehensive approach to high-quality antepartum, intrapartum and post-partum care.

4.2.11 Conclusion

Treatment of the pregnant or postpartum patient who meets case definition for EVD is controversial and is often an ethically charged debate due to the overlap in clinical presentation of EVD and obstetric complications. The 2013–2015 West Africa Ebola epidemic illuminated the desperate need for adequate preparedness for infectious disease outbreaks in the obstetric population, with dedicated protocols, adequate training, and unique considerations for these extremely vulnerable patients. The safety and protection of the healthcare worker must be balanced with the commitment to deliver the highest degree of quality clinical intervention possible for the pregnant patient, with the theoretical risk of transmission of EVD incorporated into every aspect of clinical and care management.

We recommend that the lessons learned from the 2013–2015 West African epidemic, where an unknown, yet unfathomable number of EVD negative women and infants lost their lives in Ebola Treatment Centers due to inadequate obstetric care, be considered in the development of all future emergency preparedness and response protocols. With committed partners implementing informed protocols, safe and high-quality maternal/child health can and should be prioritized in the midst of an emergency.

4.3   Ebola Virus Disease in Children

Felicity Fitzgerald

4.3.1 Epidemiology

Nearly 8000 children were confirmed or suspected to be infected with EVD during the West African 2013–2015 outbreak, just under 25% of the total [74]. It appears that in this as with previous outbreaks, confirmed diagnoses were fewer in children than adults [75–77]. The reasons underlying this apparent sparing of children are poorly understood. Firstly, it could be that diagnoses are being missed, due to under-reporting or poor diagnostic sensitivity in children [77, 78]. It is possible that against a background of high infant mortality as with most countries affected by EVD, and fear of seeking health care during an outbreak parents did not bring their unwell children for testing [78]. Alternatively, it may be that the diagnostic tests such as both PCR (polymerase chain reaction for EVD DNA or serological testing) were less sensitive for small children [75]. For example, small children, particularly infants, are often challenging to take blood samples from so it may be that smaller samples were sent, or that less sensitive mouth swabs were used for PCR as an alternative early in the outbreak [79].

However, the difference may be true biological sparing of children, where children are either less exposed or less vulnerable to exposure, or finally that children were more likely to have asymptomatic infection. From Glynn et al.'s study of seroprevalence of EVD Immunoglobulin G (IgG, evidence of previous infection) in households of EVD survivors, they found no evidence of asymptomatic infections in children under 12 years of age, and a slight excess of symptomatic undiagnosed infections in younger children: i.e. it appears that younger children were slightly more likely not to be taken to hospital despite symptoms [80]. Bower et al. investigated age-specific attack rates in the same cohort of EVD survivor households in Sierra Leone, and found that after adjustment for exposure type, children and adolescents aged 5–19 years were less vulnerable to infection than either younger children or adults [81].

Therefore it appears from available evidence that in the West African EVD outbreak at least, the apparent sparing of children was due to a combination of younger children not being brought for medical attention and true biological sparing in older children and adolescents, the mechanisms for which remain to be explained.

Routes of exposure are similar to adults with the additional exposure of breast milk for infants, and vertical transmission from mother to neonate (see obstetric chapter). It appears that close contact with a sick mother/primary caregiver is a key risk factor in children over and above other household or community exposures [82]. Regarding breast feeding, Ebola virus has been detected in breast milk up to 9 months post-infection [83–85]. Indeed, in one case, investigation into the death of a 9-month old infant from EVD led to the discovery of Ebola virus in the mother's breast milk although the mother had had no preceding symptoms [86]. However, Bower et al.'s study of 77 mother-child pairs (children aged < 2 years) found no excess risk from breast feeding over and above contact with a sick mother [82]. Contact with an EVD-infected mother was by far the greatest risk for the child, risk ratio (RR) compared with infections in the same household excluding mother 7.5, 95% confidence interval (CI) 1.9–28.9, $p < 0.001$) [82]. Interestingly, household crowding and sanitation had little impact on transmission risk, and none of the children included had contact with a dead body, indicating close proximity to the mother/primary caregiver as the most important causal factor in acquisition of EVD. The authors therefore agree with current WHO guidelines that asymptomatic infants and children should be separated from infected mothers to limit onward transmission [82, 87].

Regarding children themselves as sources of the virus, the data is conflicting. One modelling study based on data from 200 burials indicated that children might be "super-spreaders" of the virus, but this has not been substantiated by epidemiological data from Liberia or Sierra Leone [88]. In Liberia, a contact tracing study showed no difference between children and adults in terms of transmission, and a study of transmission chains indicated that children were less likely to pass on the virus [89, 90]. This was mirrored in Sierra Leone where children were more likely to be infected in later generations within households, rather than being the primary source within a household [91]. It seems likely therefore that children may be less, rather than more likely to transmit the infection compared to adults.

In terms of mortality, infants are the most vulnerable, with case fatality rates (CFRs) varying between 70 and 90% [92–95]. The prognosis improves with age, such that mid-late teenagers have amongst the lowest case fatality rates. Table 4.11 shows CFRs for children in studies from both the West African and prior outbreaks by age. Risk factors for mortality are discussed further below.

4.3.2 Presentation

EVD is notoriously non-specific in presentation, particularly in children. Indeed, even fever which was key to the WHO clinical case definition in the West African outbreak was absent in 20–25% of cases in three studies [94, 96–98]. In most studies from the West African and previous outbreaks, features in children have included (in order of frequency) fever (71–99%), fatigue/weakness (64–80%), appetite loss (60–79%), vomiting (28–62%) and diarrhoea (43–60%) [94–97, 99]. Abdominal, muscle, joint and chest pain as well as headaches have been reported in 29–70%, although in younger children pain is difficult to localise and so recording of pain from various body sites has been compounded into the symptom of generalised distress, seen in 64% of a younger cohort [94]. Conjunctivitis was recorded in 13–22% and hiccoughs in 7–12% [94, 95, 99, 100]. Difficulty breathing and swallowing were seen approximately 13–20% of patients [96, 97, 99]. Bleeding from various body sites tended to be rarer in the West African outbreak in children than previous outbreaks (1–10% compared with 20%) [95, 97, 99, 101], although two cohort studies recorded bleeding in 15% [94, 96], and a large study of Guinean children recorded bleeding in 24% [100]. Interestingly, one younger cohort (children aged up to 5 years) recorded cough in up to 54% of children, although this was less frequently seen in other cohorts [94].

Blood tests have revealed dramatic leucocytosis, deranged liver and renal function alongside raised inflammatory markers (e.g. C-reactive protein) particularly in children who died [11, 95]. Hypoglycaemia, often severe, was common among both children who died (55%) and those who survived (30%) in one cohort [95]. More detailed description of electrolyte and haematological disturbances over the course of disease can be seen in Fig. 4.1 [102].

4.3.3 Disease Progression

The mean duration of incubation of EVD is shortest in younger children: estimated to be 1 week in children <1 year compared to 9.8 days in children aged 10–15 years [99]. Similarly, duration from symptom onset to death is shortest in younger children- under 6 days in those <1 year, compared with nearly 9 days in those aged 10–15 years [99]. However, care must be taken with these estimates as many children were admitted unaccompanied to treatment facilities, so data regarding

Table 4.11 Showing case fatality rates by age from West African and Ugandan outbreaks ordered by study size

Study	Country	Age range (years)	Number within age range	Case fatality rate (%)
Garske et al.[a]	Guinea, Liberia and Sierra Leone	<1	125	80
		1–4	460	78
		5–10	527	59
		11–15	626	50
Cherif[b]	Guinea	1–4	211	83
		5–9	136	65
		10–15	131	49
Fitzgerald et al.[c]	Sierra Leone	<1	20	70
		1–4	84	65
		5–10	84	94
		11–12	94	47
Smit et al.[d]	Sierra Leone and Liberia	0–4	44	89
		5–9	30	43
		10–14	32	41
Shah et al.[e]	Sierra Leone	0–2	34	77
		2–5	57	46
McElroy et al.[f]	Uganda	0–5	13	77
		6–15	16	38
Damkjaer et al.[g]	Sierra Leone	0–18	33	42
Mupere et al.[h]	Uganda	0–17	20	40

[a]Garske T, Cori A, Ariyarajah A, et al. Heterogeneities in the case fatality ratio in the West African Ebola outbreak 2013–2016. Philos Trans R Soc Lond B Biol Sci 2017; 372(1721)

[b]Cherif MS, Koonrungsesomboon N, Kasse D, et al. Ebola virus disease in children during the 2014–2015 epidemic in Guinea: a nationwide cohort study. European journal of pediatrics 2017; 176(6): 791–6

[c]Fitzgerald F, Naveed A, Wing K, et al. Ebola Virus Disease in Children, Sierra Leone, 2014–2015. Emerging infectious diseases 2016; 22(10): 1769–77

[d]Smit MA, Michelow IC, Glavis-Bloom J, Wolfman V, Levine AC. Characteristics and Outcomes of Pediatric Patients With Ebola Virus Disease Admitted to Treatment Units in Liberia and Sierra Leone: A Retrospective Cohort Study. Clin Infect Dis 2017; 64(3): 243–9

[e]Shah T, Greig J, van der Plas LM, et al. Inpatient signs and symptoms and factors associated with death in children aged 5 years and younger admitted to two Ebola management centres in Sierra Leone, 2014: a retrospective cohort study. Lancet Glob Health 2016; 4(7): e495–501

[f]McElroy AK, Erickson BR, Flietstra TD, et al. Biomarker correlates of survival in pediatric patients with Ebola virus disease. Emerging infectious diseases 2014; 20(10): 1683–90

[g]Damkjaer M, Rudolf F, Mishra S, Young A, Storgaard M. Clinical Features and Outcome of Ebola Virus Disease in Pediatric Patients: A Retrospective Case Series. The Journal of pediatrics 2017; 182: 378–81 e1

[h]Mupere E, Kaducu OF, Yoti Z. Ebola haemorrhagic fever among hospitalised children and adolescents in northern Uganda: epidemiologic and clinical observations. Afr Health Sci 2001; 1 (2): 60–5

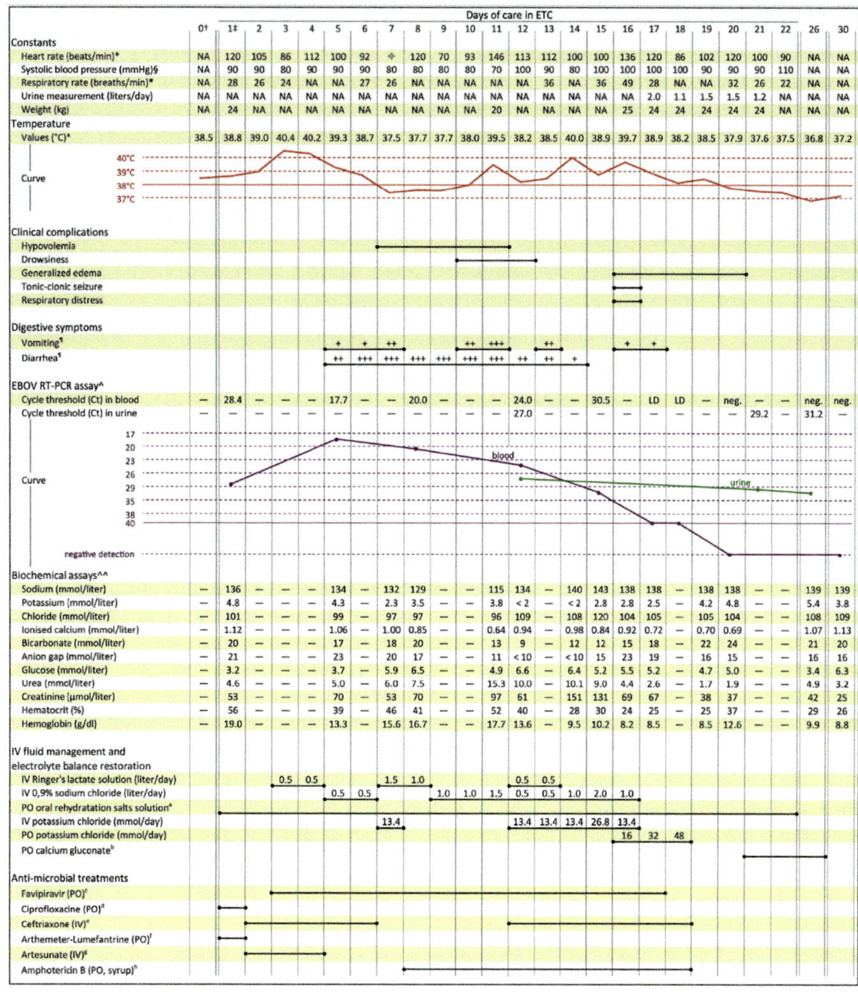

Days of care in ETC

	0†	1‡	2	3	4	5	6	7	8	9	10	11	12	13	14	15	16	17	18	19	20	21	22	26	30
Constants																									
Heart rate (beats/min)*	NA	120	105	86	112	100	92	♦	120	70	93	146	113	112	100	100	136	120	86	102	120	100	90	NA	NA
Systolic blood pressure (mmHg)§	NA	90	90	80	90	90	90	80	80	80	80	70	100	90	80	100	100	100	100	90	90	90	110	NA	NA
Respiratory rate (breaths/min)*	NA	28	26	24	NA	NA	27	26	NA	NA	NA	NA	NA	36	NA	36	49	28	NA	NA	32	26	22	NA	NA
Urine measurement (liters/day)	NA	NA	NA	NA	NA	NA	NA	NA	NA	NA	NA	NA	NA	NA	NA	NA	NA	2.0	1.1	1.5	1.5	1.2	NA	NA	NA
Weight (kg)	NA	24	NA	NA	NA	NA	NA	NA	NA	NA	NA	20	NA	NA	NA	NA	25	24	24	24	24	24	NA	NA	NA
Temperature																									
Values (°C)*	38.5	38.8	39.0	40.4	40.2	39.3	38.7	37.5	37.7	37.7	38.0	39.5	38.2	38.5	40.0	38.9	39.7	38.9	38.2	38.5	37.9	37.6	37.5	36.8	37.2
Digestive symptoms																									
Vomiting¶						+	+	++			++	+++		++				+	+						
Diarrhea¶						++	+++	+++	+++	+++	+++	+++	++	++	+										
EBOV RT-PCR assay^																									
Cycle threshold (Ct) in blood	—	28.4	—	—	—	17.7	—	—	20.0	—	—	—	24.0	—	—	30.5	—	LD	LD	—	neg.	—	—	neg.	neg.
Cycle threshold (Ct) in urine	—	—	—	—	—	—	—	—	—	—	—	—	27.0	—	—	—	—	—	—	—	—	29.2	—	31.2	—
Biochemical assays^^																									
Sodium (mmol/liter)	—	136	—	—	—	134	—	132	129	—	—	115	134	—	140	143	138	138	—	138	138	—	—	139	139
Potassium (mmol/liter)	—	4.8	—	—	—	4.3	—	2.3	3.5	—	—	3.8	<2	—	<2	2.8	2.8	2.5	—	4.2	4.8	—	—	5.4	3.8
Chloride (mmol/liter)	—	101	—	—	—	99	—	97	97	—	—	96	109	—	108	120	104	105	—	105	104	—	—	108	109
Ionised calcium (mmol/liter)	—	1.12	—	—	—	1.06	—	1.00	0.85	—	—	0.64	0.94	—	0.98	0.84	0.92	0.72	—	0.70	0.69	—	—	1.07	1.13
Bicarbonate (mmol/liter)	—	20	—	—	—	17	—	18	20	—	—	13	9	—	12	12	15	18	—	22	24	—	—	21	20
Anion gap (mmol/liter)	—	21	—	—	—	23	—	20	17	—	—	11	<10	—	<10	15	23	19	—	16	15	—	—	16	16
Glucose (mmol/liter)	—	3.2	—	—	—	3.7	—	5.9	6.5	—	—	4.9	6.6	—	6.4	5.2	5.5	5.2	—	4.7	5.0	—	—	3.4	6.3
Urea (mmol/liter)	—	4.6	—	—	—	5.0	—	6.0	7.5	—	—	15.3	10.0	—	10.1	9.0	4.4	2.6	—	1.7	1.9	—	—	4.9	3.2
Creatinine (μmol/liter)	—	53	—	—	—	70	—	53	70	—	—	97	61	—	151	131	69	67	—	38	37	—	—	42	25
Hematocrit (%)	—	56	—	—	—	39	—	46	41	—	—	52	40	—	28	30	24	25	—	25	37	—	—	29	26
Hemoglobin (g/dl)	—	19.0	—	—	—	13.3	—	15.6	16.7	—	—	17.7	13.6	—	9.5	10.2	8.2	8.5	—	8.5	12.6	—	—	9.9	8.8

IV fluid management and electrolyte balance restoration (printed values, by day of care):

	2	3	4	5	6	7	8	9	10	11	12	13	14	15	16	17	18	19
IV Ringer's lactate solution (liter/day)	0.5	0.5			1.5	1.0						0.5	0.5					
IV 0,9% sodium chloride (liter/day)			0.5	0.5				1.0	1.0	1.5	0.5	0.5	1.0	2.0	1.0			
IV potassium chloride (mmol/day)					13.4						13.4	13.4	13.4	26.8	13.4			
PO potassium chloride (mmol/day)																16	32	48

The following rows of the chart are shown as horizontal duration bars (and, where noted, as intermittent symptom/dosing markers) rather than as printed values; their approximate spans in days of care are:

- Clinical complications — Hypovolemia (≈ days 7–11); Drowsiness (≈ days 9–13); Generalized edema (≈ days 16–20); Tonic-clonic seizure (≈ days 16–17); Respiratory distress (≈ days 16–17).
- PO oral rehydratation salts solution^a (≈ days 2–17); PO calcium gluconate^b (≈ days 21–30).
- Anti-microbial treatments — Favipiravir (PO)^c (≈ days 5–14); Ciprofloxacine (PO)^d (≈ days 1–2); Ceftriaxone (IV)^e (≈ days 2–8); Arthemeter-Lumefantrine (PO)^f (≈ day 1); Artesunate (IV)^g (≈ day 1); Amphotericin B (PO, syrup)^h (≈ days 7–16).

Fig. 4.1 Clinical and biological outcomes and main treatments administered to a 6 year old with EVD during care in an Ebola treatment centre [102]. *NA* not available, *LD* limit of detection. †December 25th, 2014. ‡December 26th, 2014. *Highest measurement of the day. §Lowest measurement of the day. Pulse was not perceived because of hemodynamic disorder. ¶Gastrointestinal losses were quantified at each visit of the medical team (four times per day): severe (+++), moderate (+ +), or mild (+). Severe losses: more than eight liquid stools per day (or more than four bouts of vomiting daily). Moderate losses: more than four liquid stools (or more than two bouts of vomiting per day). Mild losses: at least one liquid stool per day (or at least one bout of vomiting per day). ^Biomolecular tests employed RealStar Filovirus RT-PCR Kit 1.0 (Altona Diagnostics GmbH, Hamburg, Germany). ^^Biochemical assays employed i-STAT CHEM8+ cartridges (Abbott Point of Care Inc., Princeton, New Jersey, USA). (a) Between 0.5 and 1.0 L per day, at will. (b) 3.5 mmol per day. (c) Loading dose of 3000 mg on the first day (H0: 1200 mg, H8: 1200 mg, H16: 600 mg) then 1200 mg (600 mg twice a day) on the following days. (d) 250 mg twice per day. (e) 1 g once a day. (f) 40 mg/240 mg twice a day. (g) 120 mg (H0 60 mg, H12 60 mg) on the first day, then 60 mg per day on the following days. (h) 500 mg twice a day. Reproduced under Creative Commons Licence from Pallich et al. [102]

symptom duration may be unreliable particularly in younger children. Duration between symptom onset to attendance was 4 days [94]. Time from presentation at a treatment facility to death can also be short (a median of 3 days in one study) so there is a small window for intervention [95].

Features at admission that are consistently associated independently with mortality across several studies include younger age and a high viral load (low cycle threshold (CT) with a viral polymerase chain reaction) as with mixed age cohorts [92, 95–97, 99, 100, 103, 104]. Shah et al. record a hazard ratio of 9.2 (95% confidence interval 3.8–22.5) for death with a CT value <25 at admission [94]. However, in interpreting viral CT values, discrepancies between laboratories and assays used should be borne in mind, as there is currently no universally used assay. Bleeding at admission and diarrhoea have also been reported to be associated with an increased risk of death [95, 97]. Finally, in mixed age cohorts, concomitant infection with malaria has also been significantly associated with mortality [104].

During admission, development of bleeding, shortness of breath and diarrhoea at any point are all independently associated with mortality, as well as tachycardia within the first week of admission [103]. Dysphagia was more common those who died and in Shah et al.'s younger cohort (children under 5 years), hiccoughs, confusion and bleeding were only present in children who died [94]. For those patients who recovered, there was a period of defervescence over 7–10 days [103]. No children died after day 13 of admission in one study [96]. Median duration of admission those who survived varies between 2 and 3 weeks [95, 97]. Palich et al. have thoroughly documented the progress of a 6 year old with severe EVD managed at a treatment centre with facilities of laboratory monitoring and intravenous fluid and electrolyte replacement. Figure 4.1 summarises the clinical features and interventions received for this child. Timing of gastrointestinal and neurological symptoms can be seen along with electrolyte disturbances and interventions.

Late complications during admission include tonic-clonic seizures, which can be prolonged and severe [95, 102]. In at least two cases, children were left with severe disabilities after a prolonged seizure including blindness and paraplegia [95] (Howlett et al. in review). In Bower et al.'s study of late deaths within a cohort of 151 EVD survivors, one 6 year old and one 17 year old died after discharge from an Ebola treatment centre [92]. The 6 year old had symptoms consistent with tuberculosis, although also had a post-mortem mouth swab that was borderline positive for Ebola virus [92]. The 17 year old died 5 weeks after discharge with weight loss, night sweats and dysphagia with a post-mortem swab negative for Ebola virus [92]. It appears that true recrudescence of virus as has been seen in adult patients may be very rare although possible [18, 105].

4.3.4 Suspect Ebola Virus Disease Case Definition in Children

As can be seen by both the breadth of symptoms exhibited by children with EVD, the variations in frequency of symptoms between cohorts even within the West African outbreak, and the non-specificity of these symptoms against a backdrop of high

Table 4.12 Scores for each of the variable included in the Paediatric Ebola Predictive score

Variable	Coefficient (95% CI) from multivariable model	p-value	Integer score value
Positive contact	2.21 (1.58–2.83)	<0.001	+2
Conjunctivitis	1.34 (0.62–2.05)	<0.001	+2
Age 2+ years	1.06 (0.37–1.75)	0.003	+2
Fever	0.99 (−0.66–2.63)	0.241	+1
Anorexia	0.59 (−0.18–1.35)	0.133	+1
Male gender	0.49 (−0.11–1.08)	0.111	+1
Abdominal pain	0.42 (−0.23–1.08)	0.205	+1
Diarrhoea	0.40 (−0.21–1.01)	0.197	+1
Difficulty breathing	−0.57 (−1.39–0.24)	0.168	−1
Difficulty swallowing	−0.59 (−1.39–0.19)	0.138	−1
Headache	−0.63 (−1.29–0.35)	0.063	−1
Skin rash	−1.00 (−2.13–0.14)	0.085	−2

malaria prevalence and other childhood illnesses, the clinical case definition for EVD is key. A sensitive and specific clinical case definition for EVD in children would allow rapid access to appropriate treatment for children with EVD, crucial when early mortality is so high, and also protect children without EVD (with a different illness) from exposure to EVD while awaiting laboratory test confirmation. To date there has been only one attempt at deriving a paediatric-specific case definition using multicentre data on over 1000 children from the West African outbreak, and this case definition has not yet been validated on wider datasets [106]. It also does not include malaria rapid diagnostic test results which have been shown to be an important discriminator in a mixed age study [107]. However, these limitations aside, the paediatric Ebola predictive score (PEP) score derived had excellent discrimination (area under receiver operating characteristics curve (AUROC) = 0.8). The scores for each clinical feature within the score are shown in Table 4.12, along with the coefficient from the multivariable model used to derive them and associated p values. As the score has not yet been externally validated, it is presented here for information rather than recommendation for use. It is envisaged that the PEP score could be used together with EVD rapid diagnostic tests, several of which were trialled during the outbreak, to expedite rapid accurate diagnosis of EVD [108, 109].

4.3.5 Management

In the stressful environment of an EVD outbreak, the needs of children can be overlooked. Certain considerations should be planned for in advance to ensure adequate care for children. These considerations include both clinical, nutritional and pastoral care of children admitted with EVD.

4.3.5.1 Assessment and Investigations

Initial assessment for emergency clinical signs should take place using the Emergency Triage and Assessment Treatment (ETAT) algorithm for children [110]. Assessment for dehydration is key. Weighing can be a challenge in facilities with limited resources, so we would suggest using a set of scales within a ziplock plastic bag (or similar) to promote easy cleaning, prevent contamination and prolong the lifespan of the scales when being cleaned with high concentration chlorine. Signs of severe or moderate dehydration should be assessed clinically on admission and at least daily (preferably more frequently during the gastrointestinal phases of disease) during admission as per guidelines in the WHO pocket book of hospital care for children [110]. Pain and distress is common in children with EVD and should be specifically assessed for and treated both at initial admission and on reassessment.

Blood should be taken for EVD PCR as per WHO recommendations for exposure prone procedures, ideally by at least two staff [9]. This is particularly important for children who need to be held still during phlebotomy. A malarial RDT should be performed if possible in high prevalence areas, as coinfection is not infrequent [107, 111]. A glucose test should also be performed as a priority if feasible.

4.3.5.2 Clinical Management

Any initial emergency clinical features noted during ETAT assessment should be managed as per the relevant treatment algorithm. Placing a laminated version of the algorithm on the wall in triage will prove useful for treating clinicians. Fever should be managed with antipyretics dosed as per weight/age, and again a laminated list of age/weight appropriate commonly used medications will prove useful for clinical staff both in and outside the "Red Zone".

Adequate hydration is a mainstay of therapy for those with both mild and severe EVD. All children should be offered oral rehydration solution and be supported to drink it, either by a caregiver or a dedicated staff member as they may be too weak to drink it themselves [112]. If the child is severely dehydrated or malnourished, or unable to tolerate oral fluids, intravenous fluid resuscitation and maintenance will be needed. This should be carried out according to the presence/absence of shock, severe malnutrition and severe anaemia according to the detailed protocols in the WHO Pocket guide to Clinical Management of patients with viral haemorrhagic fever. These include recommendations for the volume and rate of fluid replacement and are not reproduced here in the interests of space, but we recommend that laminated versions of the fluid replacement protocols be available both in the "Red Zone" and outside where fluids and medications are prepared and prescribed [79]. In brief, the three signs of shock considered are: cold extremities, weak **AND** fast pulse and a capillary refill time of over 3 s; severe anaemia is diagnosed with a haematocrit <15 or a haemoglobin less than 5 g/dL; and severe acute malnutrition is diagnosed with a MUAC of <115 mm [79].

Although in the context of large gastrointestinal fluid losses, the benefits of intravenous fluid resuscitation is unequivocal and recommended by the WHO, fluid resuscitation should ideally be carried out with input/output monitoring and using a paediatric giving set [79]. Aggressive fluid resuscitation in African children with signs of severe infection (excluding gastrointestinal infections) is not proven to be safe, so every effort should be made to monitor both the volume of fluid given and the clinical impact on the child [113]. Ongoing assessment of hydration status is therefore key, and should be planned for in daily staffing allocations. If possible, electrolytes should also be monitored and abnormalities corrected, as large derangements in sodium, potassium and calcium have all been seen in children with EVD [95, 102]. Hypoglycaemia has been demonstrated to be common in children with EVD and should be monitored for as a priority if feasible, but if monitoring is not feasible, glucose should be given intravenously empirically in the case of seizure, coma or lethargy [95]. Five percent dextrose should be used in all maintenance fluids [79]. For persistent vomiting, ondansetron (or if unavailable, promethazine, though with care to monitor for extra pyramidal side effects) may limit symptoms and permit oral intake [79].

If a malarial RDT is not available or the result is positive, children should receive a weight- or age- appropriate dose of artesunate combination therapy (ACT) or intravenous/intramuscular artesunate if there are signs of severe malaria for at least hours followed by a 3 day course of oral ACT [79].

Owing to the overlap of symptoms of EVD with sepsis, it is both the WHO and our recommendation that all children under 5 years of age admitted with suspect EVD should be treated with empirical antibiotics on admission. WHO guidelines are that all under 5 s should receive intravenous or intramuscular broad spectrum antibiotics (e.g. ceftriaxone), although evidence is limited as to whether in relatively well younger children parentral antimicrobials will provide a benefit over enteral [95]. For older children, the decision to start antibiotics lies with the treating clinician as evidence is lacking, but the local prevalence of other common childhood illnesses such as pneumonia and gastroenteritis should be taken into consideration. If used locally, the empirical treatment guidelines laid out in the WHO Pocket book of Hospital Care for Children and the Integrated Management of Neonatal and Child-hood illness should be utilised [110]. All those receiving antiretroviral or anti-tuberculous therapy should continue it, and restart as soon as possible if treatment is interrupted.

Pain and distress has been noted frequently in children with EVD (see section on clinical features at presentation and disease progression) and should be managed expectantly with age/weight appropriate doses of simple analgesia (paracetamol), or opiates (tramadol followed by morphine if available) if there is ongoing pain or distress in younger children. Dysphagia has been noted to be associated with retrosternal chest pain in adults, and should be managed in children with

Management of bleeding can include transfusion of packed red cells or components (e.g. fresh frozen plasma if required), and particularly for malnourished children supplementation of vitamin K (enterally or parentrally) should be considered. However evidence is limited, particularly as regards antifibrinolytics [79].

Other empirical management for shortnesss of breath (oxygen via nasal cannulae if available); seizures (glucose if hypoglycaemic or monitoring unavailable and benzodiazepines); confusion (reassurance, possibly sedation if agitation severe) should be given as needed/available depending on context. In the unfamiliar and frightening environment of the treatment centre, reassurance from staff or other caregivers cannot be overemphasised as a crucial therapeutic intervention.

4.3.5.3 Nutrition

Given the gastrointestinal symptoms of EVD and the background prevalence of malnutrition in the countries EVD has affected, assessment for malnutrition at admission is advisable. WHO guidelines are that a mid-upper arm circumference (MUAC) should be checked and oedema assessed for at a minimum if it not feasible to assess anthropometry in more detail [87]. As weight of children is key both in assessing hydration and nutritional status, we would suggest checking weight on admission in addition to MUAC and presence/absence of oedema. As above, scales could be kept within a ziplock plastic bag to protect from high concentration chlorine and ease of decontamination after use.

If severe acute malnutrition is present, fluid resuscitation should be given according to WHO guidelines using smaller volumes and ReSoMal™ as opposed to standard oral rehydration salts [79].

For asymptomatic breastfed infants whose mother is unwell with EVD, current recommendations based on available evidence are to separate the mother and infant and use replacement ready to use infant formula [82, 87]. If the infant also has signs/symptoms of EVD, or a confirmed diagnosis, the current guidelines are that the benefits of continuing breastfeeding are likely to outweigh the risks and so breastfeeding should be continued if the mother is well enough to do so [87].

4.3.5.4 Pastoral Care

The challenges of caring for unwell, frightened children within a "Red Zone" have been well documented during the West African outbreak [94, 96, 112, 114, 115]. This problem is amplified as many children were admitted unaccompanied by a caregiver (up to 40% in one study [95]), either because family members had already succumbed to disease or because fear of nosocomial infection prevented either relatives from wishing to enter the "Red Zone" or unit policies forbade the admission of asymptomatic caregivers. Apart from the risk of nosocomial infection between suspect EVD patients posed by unaccompanied children (who were difficult to keep in their allocated bed space); there were also the hazards of sharps bins and high concentration chlorine within the Red Zone. Units developed different practices to mitigate these risks. One treatment centre had a nursing round specifically dedicated to paediatric care 8 times a day, to manage intravenous infusions and ensure that unaccompanied small children were fed 8 times daily with therapeutic milk [96].

Other units started to develop protocols to employ survivors (believed to be at low risk of re-infection with EVD) to care for unaccompanied children, but none, to our knowledge put the protocols into practice before the end of the West African outbreak. The possibility of either dedicated paediatric clinicians or a specific paediatric area with risks e.g. sharps and chlorine buckets minimised have both been discussed, and were put into limited practice towards the end of the West African outbreak. One study collected data post-discharge from caregivers admitted to Ebola Holding Units with children who subsequently tested negative for EVD [116]. It was possible to contact approximately 25% of caregivers admitted, and of those 125 contacted, none were subsequently readmitted with EVD. For the overwhelming majority of those who it was not possible to contact, this was due to a lack of contact details. This indicates that it may not be as dangerous as anticipated to admit asymptomatic caregivers with their children, although this should be done in the context of very clear instructions about hand hygiene and strict isolation between patients.

In the event that admitting caregivers with children is not possible or not felt to be safe, alternative mechanisms for managing the pastoral care needs of children have included using stretches of clear plastic sheeting between "Red" and "Green" Zones to enable frequent communication with staff/relatives outside with children inside (Heldermann T., personal communication); or frequent ward rounds of dedicated clinical staff as above.

Post discharge, consideration should be given to the mental health and developmental impact on children who, though they may have made a full physical recovery, have endured the traumatic experience of disease, separation from family, treatment within a facility and often witnessing sickness and death of parents and other close family members [84].

Finally, the long term plight of Ebola orphans should be considered. The West African outbreak resulted in an estimated 9600 Ebola orphans across the three countries which bore the brunt [117]. Although these children comprise a small proportion of the total number of orphans across the three countries, (1.4% of 711,600), whether they themselves were infected with EVD or not, they are likely to be the victims of considerable stigma and may not be welcomed into fostering families as other orphans might be [117].

References

1. Sprecher A. Filovirus haemorrhagic fever guidelines. Brussels: Médecins Sans Frontières; 2013.
2. O'Shea MK, Clay KA, Craig DG, et al. Diagnosis of febrile illnesses other than Ebola virus disease at an Ebola treatment unit in Sierra Leone. Clin Infect Dis. 2015;61(5):795–8.
3. WHO. Clinical management of patients with viral haemhorrhagic fever. Geneva: WHO; 2016.
4. Schieffelin JS, Shaffer JG, Goba A, et al. Clinical illness and outcomes in patients with Ebola in Sierra Leone. N Engl J Med. 2014;371:2092–100.
5. Dallatomasina S, Crestani R, Sylvester Squire J, et al. Ebola outbreak in rural West Africa: epidemiology, clinical features and outcomes. Trop Med Int Health. 2015;20(4):448–54.
6. Lado M, Walker NF, Baker P, et al. Clinical features of patients isolated for suspected Ebola virus disease at Connaught Hospital, Freetown, Sierra Leone: a retrospective cohort study. Lancet Infect Dis. 2015;15(9):1024–33.
7. Bah EI, Lamah MC, Fletcher T, et al. Clinical presentation of patients with Ebola virus disease in Conakry, Guinea. N Engl J Med. 2015;372(1):40–7.

8. Ansumana R, Jacobsen KH, Sahr F, et al. Ebola in Freetown area, Sierra Leone—a case study of 581 patients. N Engl J Med. 2015;372(6):587–8.

9. Chertow DS, Kleine C, Edwards JK, Scaini R, Giuliani R, Sprecher A. Ebola virus disease in West Africa—clinical manifestations and management. N Engl J Med. 2014;371(22):2054–7.

10. Qin E, Bi J, Zhao M, et al. Clinical features of patients with Ebola virus disease in Sierra Leone. Clin Infect Dis. 2015;61(4):491–5.

11. Hunt L, Gupta-Wright A, Simms V, et al. Clinical presentation, biochemical, and haematological parameters and their association with outcome in patients with Ebola virus disease: an observational cohort study. Lancet Infect Dis. 2015;15(11):1292–9.

12. Clay KA, Johnston AM, Moore A, O'Shea MK. Targeted electrolyte replacement in patients with Ebola virus disease. Clin Infect Dis. 2015;61(6):1030–1.

13. O'Shea MK, Clay KA, Craig DG, et al. A health care worker with Ebola virus disease and adverse prognostic factors treated in Sierra Leone. Am J Trop Med Hyg. 2016;94(4):829–32.

14. Nicholson-Roberts T, Fletcher T, Rees P, et al. Ebola virus disease managed with blood product replacement and point of care tests in Sierra Leone. QJM. 2015;108(7):571–2.

15. Uyeki TM, Mehta AK, Davey RT Jr, et al. Clinical management of Ebola virus disease in the United States and Europe. N Engl J Med. 2016;374(7):636–46.

16. Brown C, Krueuls B, Baker P, Baker T, Boyles T, Lado M, Johnson O. Ebola and provision of critical care. Lancet. 2015;385(9976):1392.

17. Chertow DS, Uyeki TM, DuPont HL. Loperamide therapy for voluminous diarrhea in Ebola virus disease. J Infect Dis. 2015;211(7):1036–7.

18. Howlett P, Brown C, Helderman T, et al. Ebola virus disease complicated by late-onset encephalitis and polyarthritis, Sierra Leone. Emerg Infect Dis. 2016;22(1):150–2.

19. Sagui E, Janvier F, Baize S, et al. Severe Ebola virus infection with encephalopathy: evidence for direct virus involvement. Clin Infect Dis. 2015;61(10):1627–8.

20. Fowler RA, Fletcher T, Fischer WA 2nd, et al. Caring for critically ill patients with Ebola virus disease. Perspectives from West Africa. Am J Respir Crit Care Med. 2014;190(7):733–7.

21. Kreuels B, Witchmann D, Emmerich P, et al. A case of severe Ebola virus infection complicated by gram-negative septicemia. N Engl J Med. 2014;371(25):2394–401.

22. Perner A, Fowler RA, Bellomo R, Roberts I. Ebola care and research protocols. Intensive Care Med. 2015;41(1):111–4.

23. Murthy S, Ebola Clinical Care authors group. Ebola and provision of critical care. Lancet. 2015;385(9976):1392–3.

24. Büttner S, Koch B, Dolnik O, et al. Extracorporeal virus elimination for the treatment of severe Ebola virus disease—first experience with lectin affinity plasmapheresis. Blood Purif. 2014;38 (3–4):286–91.

25. Mupapa K, Massamba M, Kibadi K, et al. Treatment of Ebola hemorrhagic fever with blood transfusions from convalescent patients. International Scientific and Technical Committee. J Infect Dis. 1999;179(Suppl 1):S18–23.

26. Dunning J, Kennedy SB, Antierens A, et al. Experimental treatment of Ebola virus disease with Brincidofovir. PLoS One. 2016;11(9):e0162199.

27. Sissoko D, Laouenan C, Folkesson E, et al. Experimental treatment with Favipiravir for Ebola virus disease (the JIKI trial): a historically controlled, single-arm proof-of-concept trial in Guinea. PLoS Med. 2016;13(3):e1001967.

28. Bai CQ, Mu JS, Kargbo D, et al. Clinical and virological characteristics of Ebola virus disease patients treated With Favipiravir (T-705)-Sierra Leone, 2014. Clin Infect Dis. 2016;63 (10):1288–94.

29. Dunning J, Sahr F, Rojek A, et al. Experimental treatment of Ebola virus disease with TKM-130803: a single-arm phase 2 clinical trial. PLoS Med. 2016;13(4):e1001997.

30. PREVAIL II Writing Group Multi-National PREVAIL II Study Team. A randomized, controlled trial of ZMapp for Ebola virus infection. N Engl J Med. 2016;375:1448–56.

31. van Griensven J, Edwards T, de Lamballerie X, et al. Evaluation of convalescent plasma for Ebola virus disease in Guinea. N Engl J Med. 2016;374:33–42.

32. Sahr F, Ansumana R, Massaquoi TA, Idriss BR, Sesay FR, Lamin JM, et al. Evaluation of convalescent whole blood for treating Ebola Virus Disease in Freetown, Sierra Leone. J Infect. 2017;74(3):302–9.
33. Konde MK, Baker DP, Traore FA, Sow MS, Camara A, Barry AA, et al. Interferon beta-1a for the treatment of Ebola virus disease: a historically controlled, single-arm proof-of-concept trial. PLoS One 2017;12(2):e0169255.
34. Brown CS, Houlihan CF, Lado M, Mounter N, Youkee D. What do we know about controlling Ebola virus disease outbreaks? Retrovirology. 2017;8:1–12.
35. Lanini S, Zumla A, Ioannidis JP, et al. Are adaptive randomised trials or non-randomised studies the best way to address the Ebola outbreak in west Africa? Lancet Infect Dis. 2015;15 (6):738–45.
36. Thi EP, Mire CE, Lee AC, et al. Lipid nanoparticle siRNA treatment of Ebola-virus-Makona-infected nonhuman primates. Nature. 2015;521(7552):362–5.
37. Jacobs M, Aarons E, Bhagani S, et al. Post-exposure prophylaxis against Ebola virus disease with experimental antiviral agents: a case-series of health-care workers. Lancet Infect Dis. 2015;15(11):1300–4.
38. Centers for Disease Control and Prevention. Outbreaks chronology: Ebola virus disease, Ebola hemorrhagic fever. CDC; 2015. Retrieved from http://www.cdc.gov/vhf/ebola/outbreaks/history/chronology.html
39. WHO, UNFPA, UNICEF, AMDD. Monitoring emergency obstetric care, 164; 2009. Retrieved from http://apps.who.int/iris/bitstream/10665/44121/1/9789241547734_eng.pdf
40. World Health Organization. GHO, By category, Maternal mortality – data by country. Global Health Observatory data repository: maternal mortality data by country. World Health Organization; 2016a.
41. World Health Organization. WHO, Guinea. WHO. World Health Organization; 2016b. Retrieved from http://www.who.int/countries/gin/en/
42. World Health Organization. WHO, Liberia. WHO. World Health Organization; 2016c. Retrieved from http://www.who.int/countries/lbr/en/
43. World Health Organization. WHO, Sierra Leone. WHO. World Health Organization; 2016d. Retrieved from http://www.who.int/countries/sle/en/
44. Estimates by WHO, UNICEF, UNFPA, World Bank Group and the United Nations Population Division, Trends in Maternal Mortality: 1990 to 2015. Retrieved from http://apps.who.int/iris/bitstream/10665/112682/2/9789241507226_eng.pdf
45. Maternal Mortality Estimation Interagency Group. Maternal mortality 1990–2015, Sierra Leone. 2016. http://www.who.int/gho/maternal_health/countries/sle.pdf. Retrieved 2 Apr 2016.
46. World Health Organization. Density of doctors, nurses and midwives in the 49 priority countries; 2010.
47. Brolin Ribacke KJ, van Duinen AJ, Nordenstedt H, Höijer J, Molnes R, Froseth TW, et al. The impact of the West Africa Ebola outbreak on obstetric health care in Sierra Leone. PLoS One. 2016;11(2):e0150080. https://doi.org/10.1371/journal.pone.0150080
48. World Health Organization. Ebola Situation Report – 23 September 2015, Ebola; 2015. Retrieved from http://apps.who.int/ebola/current-situation/ebola-situation-report-23-september-2015
49. Evans DK, Goldstein M, Popova A. Health-care worker mortality and the legacy of the Ebola epidemic. Lancet Glob Health. 2015;3(8):e439–40. https://doi.org/10.1016/S2214-109X(15)00065-0
50. Diggins J, Mills E. The pathology of inequality: gender and Ebola in West Africa. IDS; 2015. Retrieved from http://opendocs.ids.ac.uk/opendocs/handle/123456789/5856
51. Gerntholtz L, Gibbs A, Willan S. The African Women's Protocol: bringing attention to reproductive rights and the MDGs. PLoS Med. 2011;8(4):e1000429. https://doi.org/10.1371/journal.pmed.1000429.
52. Sow, M. S., Etard, J.-F., Baize, S., Magassouba, N., Faye, O., Msellati, P., et al. New evidence of long-lasting persistence of Ebola virus genetic material in semen of survivors. J Infect Dis. 2016;jiw078. https://doi.org/10.1093/infdis/jiw078

53. Mcmichael A, Simon AK, Hollander GA. Evolution of the immune system in humans from infancy to old age. Proc Biol Sci. 2015;282(1821):20143085. https://doi.org/10.1098/rspb.2014.3085.
54. UNFPA. The state of the world's midwifery 2014; 2014. Retrieved from http://www.unfpa.org/sites/default/files/pub-pdf/EN_SoWMy2014_complete.pdf
55. WHO Library Cataloguing-in-Publication Data. Clinical management of patients with viral haemorrhagic fever. A pocket guide for front-line health workers; 2016. Retrieved from http://apps.who.int/iris/bitstream/10665/205570/1/9789241549608_eng.pdf?ua=1
56. Haaskjold YL, Bolkan HA, Krogh KØ, Jongopi J, Lundeby KM, Mellesmo S, Blomberg B. Clinical features of and risk factors for fatal Ebola virus disease, Moyamba District, Sierra Leone, December 2014–February 2015. Emerg Infect Dis. 2016;22(9):1537–44. https://doi.org/10.3201/eid2209.151621.
57. Schieffelin JS, Shaffer JG, Goba A, Gbakie M, Gire SK, Colubri A, Garry RF. Clinical illness and outcomes in patients with Ebola in Sierra Leone. N Engl J Med. 2014;371(22):2092–100. https://doi.org/10.1056/NEJMoa1411680.
58. Bower H, Grass JE, Veltus E, Brault A, Campbell S, Basile AJ, et al. Delivery of an Ebola virus-positive stillborn infant in a rural community health center, Sierra Leone, 2015. Am J Trop Med Hyg. 2016;94(2):417–9. https://doi.org/10.4269/ajtmh.15-0619.
59. Mpemba F, Kampo S, Zhang X. Towards 2015: post-partum haemorrhage in sub-Saharan Africa still on the rise. J Clin Nurs. 2014;23(5–6):774–83. https://doi.org/10.1111/jocn.12126.
60. World Health Organization. WHO | Stillbirths. 2016e. http://www.who.int/maternal_child_adolescent/epidemiology/stillbirth/en/. Retrieved 25 Sept 2016.
61. Deaver JE, Siempre Salud A, Alta C, Wayne Cohen PR, Cohen WR. Ebola virus screening during pregnancy in West Africa: unintended consequences. J Perinat Med. 2015;43(6):649–55. https://doi.org/10.1515/jpm-2015-0118.
62. Akerlund E, Prescott J, Tampellini L. Shedding of Ebola virus in an asymptomatic pregnant woman. N Engl J Med. 2015;372(25):2467–9. https://doi.org/10.1056/NEJMc1503275.
63. Caluwaerts S, Lagrou D. Guidance paper Ebola Treatment Centre (ETC): Pregnant & lactating women. Table of content. 2014. Retrieved from https://www.rcog.org.uk/globalassets/documents/news/etc-preg-guidance-paper.pdf.
64. World Health Organization. WHO/OMS. Setting up an Ebola Treatment Center (ETC); 2016e. https://extranet.who.int/ebolafmt/sites/default/files/ETC_considerations_for_set_up.pdf
65. World Health Organization. WHO model list of essential medicines 19th list WHO model list of essential medicines (April 2015) Explanatory notes. 2015c. Retrieved from http://www.who.int/medicines/publications/essentialmedicines/en/.
66. World Health Organization. Guidelines for the treatment of malaria. 3rd ed. 2015b. Retrieved from http://apps.who.int/iris/bitstream/10665/162441/1/9789241549127_eng.pdf.
67. MacLennan K, O'Brien K, Macnab WR. Core topics in obstetric anaesthesia. Cambridge University Press; 2015.
68. Dörnemann J, Burzio C, Ronsse A, Sprecher A, De Clerck H, Van Herp M, et al. First newborn baby to receive experimental therapies survives Ebola virus disease. J Infect Dis. 2017;jiw493. https://doi.org/10.1093/infdis/jiw493.
69. Nelson JM, Griese SE, Goodman AB, Peacock G. Live neonates born to mothers with Ebola virus disease: a review of the literature. J Perinatol. 2016;36(6):411–4. https://doi.org/10.1038/jp.2015.189.
70. Streifel C. How did Ebola impact maternal/child health in Liberia and Sierra Leone? Center for Strategic and International Studies; 2015. https://csis-prod.s3.amazonaws.com/s3fs-public/legacy_files/files/publication/151019_Streifel_EbolaLiberiaSierraLeone_Web.pdf
71. Kamali A, Jamieson DJ, Kpaduwa J, Schrier S, Kim M, Green NM, et al. Pregnancy, labor, and delivery after Ebola virus disease and implications for infection control in obstetric services, United States. Emerg Infect Dis. 2016;22(7):1156–61. https://doi.org/10.3201/eid2207.160269.
72. Fallah MP, et al. Pregnancy outcomes in Liberian women who conceived after recovery from Ebola virus disease. Lancet Glob Health. 2016;4(10):e678–9.

73. Centers for Disease Control and Prevention. Recommendations for breastfeeding/infant feeding in the context of Ebola virus disease; 2016. https://www.cdc.gov/vhf/ebola/hcp/recommendations-breastfeeding-infant-feeding-ebola.html
74. Garske T, Cori A, Ariyarajah A, et al. Heterogeneities in the case fatality ratio in the West African Ebola outbreak 2013–2016. Philos Trans R Soc Lond B Biol Sci. 2017; 372(1721).
75. Glynn JR. Age-specific incidence of Ebola virus disease. Lancet. 2015;386(9992):432.
76. Dowell SF, Mukunu R, Ksiazek TG, Khan AS, Rollin PE, Peters CJ. Transmission of Ebola hemorrhagic fever: a study of risk factors in family members, Kikwit, Democratic Republic of the Congo, 1995. Commission de Lutte contre les Epidemies a Kikwit. J Infect Dis. 1999;179 (Suppl 1):S87–91.
77. Dowell SF. Ebola hemorrhagic fever: why were children spared? Pediatr Infect Dis J. 1996;15 (3):189–91.
78. Helleringer S, Noymer A, Clark SJ, McCormick T. Did Ebola relatively spare children? Lancet. 2015;386(10002):1442–3.
79. World Health Organization. Clinical management of patients with viral haemorrhagic fever: a pocket guide for front-line health workers. Geneva: World Health Organisation; 2016.
80. Glynn JR, Bower H, Johnson S, et al. Asymptomatic infection and unrecognised Ebola virus disease in Ebola-affected households in Sierra Leone: a cross-sectional study using a new non-invasive assay for antibodies to Ebola virus. Lancet Infect Dis. 2017;17(6):645–53.
81. Bower H, Johnson S, Bangura MS, et al. Exposure-specific and age-specific attack rates for Ebola virus disease in Ebola-affected households, Sierra Leone. Emerg Infect Dis. 2016;22 (8):1403–11.
82. Bower H, Johnson S, Bangura MS, et al. Effects of mother's illness and breastfeeding on risk of Ebola virus disease in a cohort of very young children. PLoS Negl Trop Dis. 2016;10(4): e0004622.
83. Nordenstedt H, Bah EI, de la Vega MA, et al. Ebola virus in breast milk in an Ebola virus-positive mother with twin babies, Guinea, 2015. Emerg Infect Dis. 2016;22(4):759–60.
84. Vetter P, Kaiser L, Schibler M, Ciglenecki I, Bausch DG. Sequelae of Ebola virus disease: the emergency within the emergency. Lancet Infect Dis. 2016;16(6):e82–91.
85. World Health Organization. Clinical care for survivors of Ebola virus disease 22 January 2016. World Health Organization, Geneva; 2016.
86. Sissoko D, Keita M, Diallo B, et al. Ebola virus persistence in breast milk after no reported illness: a likely source of virus transmission from mother to child. Clin Infect Dis. 2017;64 (4):513–6.
87. WHO. Nutritional Care of children and adults with Ebola virus disease in treatment centres; 2015.
88. Lau MS, Dalziel BD, Funk S, et al. Spatial and temporal dynamics of superspreading events in the 2014–2015 West Africa Ebola epidemic. Proc Natl Acad Sci USA. 2017;114(9):2337–42.
89. Skrip LA, Fallah MP, Gaffney SG, et al. Characterizing risk of Ebola transmission based on frequency and type of case-contact exposures. Philos Trans R Soc Lond B Biol Sci. 2017; 372 (1721).
90. Lindblade KA, Kateh F, Nagbe TK, et al. Decreased Ebola transmission after rapid response to outbreaks in remote areas, Liberia, 2014. Emerg Infect Dis. 2015;21(10):1800–7.
91. Glynn JR, Bower H, Johnson S, Turay C, Sesay D, Mansaray SH, Kamara O, Kamara AJ, Bangura MS, Checchi F. Variability in intrahousehold transmission of Ebola virus, and estimation of the household secondary attack rate. J Infect Dis. 2018;217(2):232–7. https://doi.org/10.1093/infdis/jix579.
92. Bower H, Smout E, Bangura MS, et al. Deaths, late deaths, and role of infecting dose in Ebola virus disease in Sierra Leone: retrospective cohort study. BMJ. 2016;353:i2403.
93. Ebola Response WHO, Team A-AJ, Ariyarajah A, et al. Ebola virus disease among male and female persons in West Africa. N Engl J Med. 2016;374(1):96–8.
94. Shah T, Greig J, van der Plas LM, et al. Inpatient signs and symptoms and factors associated with death in children aged 5 years and younger admitted to two Ebola management centres in Sierra Leone, 2014: a retrospective cohort study. Lancet Glob Health. 2016;4(7):e495–501.

95. Fitzgerald F, Naveed A, Wing K, et al. Ebola virus disease in children, Sierra Leone, 2014–2015. Emerg Infect Dis. 2016;22(10):1769–77.
96. Damkjaer M, Rudolf F, Mishra S, Young A, Storgaard M. Clinical features and outcome of Ebola virus disease in pediatric patients: a retrospective case series. J Pediatr. 2017;182:378–81.e1.
97. Smit MA, Michelow IC, Glavis-Bloom J, Wolfman V, Levine AC. Characteristics and outcomes of pediatric patients with Ebola virus disease admitted to treatment units in Liberia and Sierra Leone: a retrospective cohort study. Clin Infect Dis. 2017;64(3):243–9.
98. World Health Organization. Clinical management of patients in the Ebola treatment Centres and other Care Centres in Sierra Leone: A Pocket Guide. Interim emergency guidelines. Sierra Leone adaptation; 2014.
99. WHO Ebola Response Team, Agua-Agum J, Ariyarajah A, et al. Ebola virus disease among children in West Africa. N Engl J Med. 2015;372(13):1274–7.
100. Cherif MS, Koonrungsesomboon N, Kasse D, et al. Ebola virus disease in children during the 2014–2015 epidemic in Guinea: a nationwide cohort study. Eur J Pediatr. 2017;176(6):791–6.
101. Mupere E, Kaducu OF, Yoti Z. Ebola haemorrhagic fever among hospitalised children and adolescents in northern Uganda: epidemiologic and clinical observations. Afr Health Sci. 2001;1(2):60–5.
102. Palich R, Gala JL, Petitjean F, et al. A 6-year-old child with severe Ebola virus disease: laboratory-guided clinical care in an Ebola treatment center in Guinea. PLoS Negl Trop Dis. 2016;10(3):e0004393.
103. Skrable K, Roshania R, Mallow M, Wolfman V, Siakor M, Levine AC. The natural history of acute Ebola virus disease among patients managed in five Ebola treatment units in West Africa: a retrospective cohort study. PLoS Negl Trop Dis. 2017;11(7):e0005700.
104. Hartley MA, Young A, Tran AM, et al. Predicting Ebola severity: a clinical prioritization score for Ebola virus disease. PLoS Negl Trop Dis. 2017;11(2):e0005265.
105. Jacobs M, Rodger A, Bell DJ, et al. Late Ebola virus relapse causing meningoencephalitis: a case report. Lancet. 2016;388(10043):498–503.
106. Fitzgerald F, Wing K, Naveed A, Gbessay M, Ross J, Checchi F, Youkee D, Jalloh MB, Baion DE, Mustapha A, Jah H, Lako S, Oza S, Boufkhed S, Feury R, Bielicki J, Williamson E, Gibb DM, Klein N, Sahr F, Yeung S. Development of a Pediatric Ebola Predictive Score, Sierra Leone. Emerg Infect Dis. 2018;24(2):311–9.
107. Hartley MA, Young A, Tran AM, et al. Predicting Ebola infection: a malaria-sensitive triage score for Ebola virus disease. PLoS Negl Trop Dis. 2017;11(2):e0005356.
108. Walker NF, Brown CS, Youkee D, et al. Evaluation of a point-of-care blood test for identification of Ebola virus disease at Ebola holding units, Western Area, Sierra Leone, January to February 2015. Euro Surveill. 2015;20(12).
109. Broadhurst MJ, Brooks TJ, Pollock NR. Diagnosis of Ebola virus disease: past, present, and future. Clin Microbiol Rev. 2016;29(4):773–93.
110. World Health Organization. Pocket book of hospital care for children: guidelines for the management of common childhood illnesses. In: . Geneva: World Health Organization. p. 2013.
111. Barry M, Traore FA, Sako FB, et al. Ebola outbreak in Conakry, Guinea: epidemiological, clinical, and outcome features. Med Mal Infect. 2014;44(11–12):491–4.
112. Fitzgerald F, Awonuga W, Shah T, Youkee D. Ebola response in Sierra Leone: the impact on children. J Infect. 2016;72:S6–S12.
113. Maitland K, Kiguli S, Opoka RO, et al. Mortality after fluid bolus in African children with severe infection. N Engl J Med. 2011;364(26):2483–95.
114. Trehan I, Kelly T, Marsh RH, George PM, Callahan CW. Moving towards a more aggressive and comprehensive model of care for children with Ebola. J Pediatr. 2016;170:28–33.e7.
115. Aswani V. Being a pediatrician in an Ebola epidemic. Pediatrics. 2015.
116. Fitzgerald F, Wing K, Naveed A, Gbessay M, Ross JCG, Checchi F, Youkee D, Jalloh MB, Baion D, Mustapha A, Jah H, Lako S, Oza S, Boufkhed S, Feury R, Bielicki J, Williamson E, Gibb DM, Klein N, Sahr F, Yeung S. Risk in the "Red Zone": outcomes for children admitted to Ebola holding units in Sierra Leone without Ebola virus disease. Clin Infect Dis. 2017;65(1):162–5. https://doi.org/10.1093/cid/cix223.
117. Evans DK, Popova A. West African Ebola crisis and orphans. Lancet. 2015;385(9972):945–6.

Diagnostics in Ebola Virus Disease

5

Colin S. Brown, Robert Shorten, and Naomi F. Walker

Contents

5.1 Introduction

Rapid identification and appropriate isolation of cases is a key process in control of any infectious disease outbreak. This process played an important part in the epidemiology of the West African Ebola Virus Disease Outbreak in that:

C. S. Brown (✉)
King's Sierra Leone Partnership, King's Centre for Global Health, King's Health Partners, King's College London, London, UK

Department of Infection, Royal Free London NHS Foundation Trust, London, UK

National Infection Service, Public Health England, London, UK
e-mail: colinbrown@doctors.net.uk

R. Shorten
Public Health Laboratory Manchester, Manchester Royal Infirmary, Manchester, UK

Department of Infection, Centre for Clinical Microbiology, University College London, London, UK

N. F. Walker
Department of Clinical Research, London School of Hygiene and Tropical Medicine, London, UK

Hospital for Tropical Diseases, University College London Hospital, London, UK

© Springer International Publishing AG, part of Springer Nature 2018   145
M. Lado (ed.), *Ebola Virus Disease*, https://doi.org/10.1007/978-3-319-94854-6_5

1. Poor access to diagnostic tests in the early months of the outbreak likely contributed to widespread transmission/dissemination
2. Improved access to diagnostic tests later in the outbreak likely enabled:
    a. Improved individual case management
    b. More efficient use of available resources (bed spaces)
    c. Reduced community and nosocomial transmission
    d. Improved provision of non-Ebola health services
    e. Eventual outbreak control

Whilst an evidence base for the impact of diagnostic testing on outbreak control is lacking, the clinical scenarios in the Box below (see Clinical scenarios: The trouble with testing) provide examples of practical problems faced when rapid accurate testing is not available.

Identification of Ebola virus in clinical samples for diagnosis has traditionally taken place in a laboratory setting, requiring Level 4 Biosafety Procedures, using PCR, a relatively technology heavy modality. In West Africa, improvements involved adapting this technology for field use, and increasing the number and capacity of local laboratories. More recently, advances in technology to develop low tech point-of-care methodologies e.g. lateral flow technology have occurred. However, at the time of publication, these were not in routine field use.

Here we summarise the key methods of EVD diagnosis, the practical challenges and report on new technologies that have the potential to impact on future clinical management.

5.2   Who to Test and How to Test?

Testing for EVD is undertaken by trained personnel using personal protective equipment (PPE), see Box 5.1, usually on a blood sample for EVD PCR. The decision to test a patient for EVD is usually based on a patient meeting a clinical "suspected case definition" (see previous discussion Chaps. 2 and 3, pp. xx). The specifics of this definition vary, and are based on local epidemiology but require regular monitoring and updating during the course of an outbreak. In a large outbreak, testing of patients who do not meet the suspect case definition may additionally be considered if an exposure prone procedure is being undertaken/underway, such as surgery or delivery, although this is not routine. If a patient dies prior to testing, and is considered a suspected case, it is possible to take an oral swab for PCR, and/or skin snips, which can stored in formalin for immunohistochemistry <6 weeks later, although caution should be taken when collecting these samples due to the high infectivity of body fluids and tissue post-mortem. Samples should ideally be stored at 4 °C (refrigerated or in a cooler box) during transport and triple contained.

5.3   Laboratory Diagnosis of EVD

Ebolavirus is a genus containing five species of Ebola virus. They are single-stranded RNA viruses containing seven structural proteins. The mainstay of EVD diagnosis has traditionally been identification of viral RNA in the blood by nucleic acid amplification test (NAAT/PCR) , using a probe/probes that typically targets the sequence of one or more of the structural proteins e.g. the nucleoprotein (NP). A number of different assays are available (see Table 5.1). This method typically has a very high sensitivity and specificity for EVD diagnosis (>99%), such that it is considered gold standard, although it is likely that sensitivity is lower in patients who have had symptoms for less than 72 h. If the blood sample is taken soon after symptom onset it is advised that a second sample be collected >72 h after symptom

Table 5.1 Examples of diagnostic nucleic acid amplification tests (NAATs)

	Assay name	Assay description and target[a]	Limit of detection[a]
Altona (Hamburg, Germany)[b]	RealStarFilovirus Screen RealStarEbolavirus	Real time NAAT targeting the L gene of all five Ebola viruses Real time NAAT for detection and differentiation of all five Ebola viruses	3.16 copies/µL 1 PFU/mL
BioMerieux (France)	BioFire Film Array Biothreat-E test	Real time NAAT for detection of ZEBOV blood and urine within approximately 1 h	600,000 PFU/mL
Cepheid (USA)	Xpert Ebola Assay	Real time NAAT for detection of ZEBOV blood and urine within approximately 2 h	232 copies/mL
Trombley assay[b]	Various	Non-commercial pan-VHF multiplex NAAT	0.001–1.0 PFU/PCR
Panning Assay[b]	Various	Non-commercial pan-Filovirus multiplex NAAT targeting L gene	10 copies per assay in 3–4 h
Roche (Switzerland) and TIB MOLBIOL GmbH (Germany)	LightMix Ebola Zaire rRt-PCR	Real time PCR targeting the L gene Up to 96 results in just over 3 h	4781 PFU/mL
Biocartis, Janssen Diagnostics and the Institute for Tropical Medicine in Antwerp (Belgium)	Idylla™ system	Real time NAAT on a fully automated molecular diagnostic platform using 0.2 mL of blood within approximately 100 min	465 PFU/mL

[a]if available in published literature or from the manufacturer. *PFU* plaque forming unit, *NAAT* nucleic acid amplification test, *PCR* polymerase chain reaction, *ZEBOV* Zaire Ebola virus
[b]Storage of reagents at −20 °C

onset to exclude EVD, if clinical suspicion is high and the initial NAAT/PCR test is negative.

NAAT, (see Table 5.1) provide a qualitative result (positive or negative), but may also give an indication of the viral load (if quantitative or semi-quantitative PCR) which can be used for monitoring.

Box 5.1 Steps in Diagnosis of Ebola Virus Disease

1. Patient is identified as a suspected case of EVD/VHF using a locally approved current "Suspect case definition" (link to page no).
2. Patient is managed in an isolation facility by staff using appropriate PPE (insert link to page no.)
3. Venous blood is collected by trained staff member wearing appropriate PPE (ideally staff members working together in pairs), into appropriate tubes and labelled with patient ID no.
4. Blood tubes are decontaminated externally for removal from isolation unit
5. Tubes are packaged (triple contained) for safe transport to the laboratory according to locally agreed policies, with patient details.
6. Tubes arrive in laboratory and are processed in designated laboratory, following safety regulations (see below).
7. NAAT/PCR assays typically involve the following steps:
   a. Inactivation of virus
   b. RNA extraction
   c. RNA amplification
   d. Concurrent positive and negative controls according to assay instructions.
8. Result available may be qualitative (positive or negative) or quantitative (allow estimation of the viral load). A result may be indeterminate if the positive or negative control did not work.
9. Results are communicated to the clinical team
10. If result is positive patient is a confirmed EVD case, if negative EVD is excluded, unless symptoms for less than 72 h—clinicians may decide to re-test for EVD at 72 h post symptom onset.
11. If the result is negative, the patient may have another infectious viral pathogen, e.g. Lassa, so consider testing for other locally relevant causes.

Regulations exist in most settings to minimise risk of contamination in the laboratory, although procedures may differ depending on the facilities/resources available. For example, in the UK laboratories consist of large, secure facilities with restricted access, carefully controlled air-handling systems with filtered air and stringent policies for control and inactivation of hazardous waste. Skilled staff undergo extensive training to work at this level of containment. The scientist is protected by a specialised suit or by manipulation of specimens exclusively within a microbiological safety cabinet. As building, commissioning, maintaining and staffing safe laboratory facilities is expensive and labour intensive, outside of an outbreak (or in a non-endemic setting) work is usually restricted national or regional centres, for example the Uganda Viral Research Institute (UVRI) in Entebbe, Uganda, which handles samples from suspected cases nationally. This typically results in a delay to obtain results once the need for testing is established, although the length of the delay will depend on a number of factors including local transport infrastructure.

Initial samples in the West African Ebola outbreak were analysed in France and Germany as there was no EVD testing available in the region at the time [1]. Expansion of laboratory facilities in the epidemic countries lagged behind transmission for most of the outbreak. Transport mechanisms were not robust and turn-around times for EVD diagnostic results were very slow, often several days to weeks at the beginning of the outbreak, until more resources were available to expand EVD testing and formal sample transport systems established. Many organisations that ran EHUs/ETCs did not have access to supportive diagnostics for non-EVD infections. That meant that patients received syndromic management for other causes of febrile illness, including malaria.

Challenges to the expansion of laboratory-based diagnostics were considerable and included the need for a reliable power supply for equipment and cold storage of molecular reagents. Air conditioning was required to prevent high environmental temperatures interfering with the optimal performance of some NAAT platforms and other point of care analysers. Like the clinical facilities, laboratories produced considerable volumes of hazardous waste which must be stored safely until appropriate inactivation and disposal can be arranged. In the latter stages of the West African EVD outbreak, with expansion of ETCs, laboratory facilities were incorporated into new centres, located in intermediate zones so they could safely receive samples from both the Red Zone of the ETC and the community. These facilities would generally not be under negative air pressure or be sealable for fumigation.

5.4 Point of Care Tests and Rapid Diagnostic Testing

Reliable point of care rapid diagnostics that do not require laboratory infrastructure would be extremely valuable in an EVD outbreak setting for reasons of speed, efficiency and cost. The term "rapid diagnostic" and "point-of-care test" have been used in different ways. Here, we reserve the use "point-of-care (POC)" rapid diagnostic to imply a test that can be performed by a minimally trained clinical worker at the bedside. Prior to the West African Outbreak, there was no commercially available rapid diagnostic test, although several tests were in development. Two published field studies reported on use of lateral flow devices, based on specific antigen-antibody binding, with good sensitivity/specificity, during the West African EVD Outbreak [2, 3]. Such tests, where the specificity is high potentially provide a rapid, low cost, simple rule-out mechanism, which could supplement clinical

Table 5.2 Examples of point of care rapid diagnostics for EVD currently available or in development

Company	Assay name	Assay description	Stage of development	Performance notes
Corgenix[a,b] (USA)	ReEBOV Antigen Rapid Test	Lateral flow device for the detection of VP40 antigen in blood in approximately 20 min	FDA approved. WHO for use in the detection of EVD where it is not possible to use a molecular assay	Sens 100% Spec 92.2% (tested against Altona NAAT) Requires storage at 4 °C
Chun-Yan Yen and colleagues	Multiplexed lateral flow assay	Silver nanoparticles conjugated to detect EBOV GP (as well as Dengue and yellow fever viruses)	Undergoing field studies	
STADApharm GmbH, developed by Senova (Germany)[a]	Ebola lateral flow test	Hand-held lateral flow device for the detection of EBOV antigens in blood and body fluids in 10 min	Commercially available	Analytical Sens CT $\leq$ 23, Spec 98%
France's Atomic Energy Commission (CEA) with Vedalab	Ebola eZYSCREEN® lateral flow assay	Hand-held lateral flow device for the detection of EBOV antigens in blood, plasma and urine in less than 15 min	Commercially available	Sens 65.3% Spec 98.9% (whole blood) Stable at 30 °C for 393 days
The United Kingdom's Defence Science and Technology Laboratory (DSTL)[b]	Lateral Flow assay	Semi-quantitative detection of undisclosed EBOV antigen using capillary blood in 20 min	Under development by BBI Solutions	Sens 100% Spec 96.6% (whole blood) Result in 20 min Does not require cold storage
SD Biosensor Inc. (Republic of Korea)[a]	SD Q Line Ebola Zaire Ag	Chromatographic immunoassay to EVOV antigens (NP, GP and VP40)		Sens 84.9% Spec 99.7% on whole blood/ plasma

(continued)

Table 5.2 (continued)

Company	Assay name	Assay description	Stage of development	Performance notes
Orasure Technologies, Inc.[a]	Ora Quick® Ebola Rapid Antigen Test kit	Immuno-chromatographic Assay to VP antigen, result in 30 min	Commercially available	Sens 84% Spec 98% – Whole blood Sens 94% Spec 100% – Cadaveric fluid

aGiven WHO Emergency Use Assessment and Listing (EUAL) approval
bField-tested during West African Ebola Outbreak
Sens Sensitivity, *Spec* specificity

screening algorithms to take the pressure off holding unit bed capacity and formal laboratory testing. For optimal benefit, these tests need to be stocked in rural health centres, in endemic areas, for use for early case detection. However, these have not to our knowledge been put to use outside a research setting. Examples are given in Table 5.2.

Other innovative rapid diagnostics, such as Gene Xpert (a type of automated PCR), potentially offer nucleic acid detection in the clinical setting, by adapting technology, to reduce the complexity, cost and expertise required for molecular diagnosis. This type of technology is likely to be of benefit in future outbreaks.

5.5 Other Tests of Infection: Serology, Antigen Testing and Viral Isolation

Serological tests are available and indicate current (IgM+) or past (IgG+) infection, usually measured by ELISA. They have not been used routinely for diagnosis of acute infections as they also require laboratory infrastructure, but are likely to be less sensitive and specific than PCR, with more variable interpretation. In past and recent outbreaks, serological tests have been used in epidemiological studies to retrospectively confirm transmission and identify asymptomatic cases. Viral antigens can also be measured by ELISA. Blood or tissue samples can also be frozen at $-20\,^{\circ}\text{C}$ for viral isolation (culture) at a later date, but this is not routinely done and is technically challenging.

5.6 Monitoring and Discharge Planning

Higher viral loads (indicated by a lower cycle threshold (CT) value on PCR) are associated with increased mortality, and so this information has prognostic value. Serial testing may be helpful to monitor the clinical condition if resources/capacity are available. Once symptoms have resolved, further sequential blood samples are tested to inform discharge planning. A negative result (failure of the viral RNA to amplify above the lower limit of detection LLOD) together with clinical resolution of symptoms, suggests the patient has recovered and is no longer infectious, with the caveat that the virus may persist in immune privileged sites. Routinely, a negative sample 72 h after resolution of fever is required for discharge. However, evidence for this policy is lacking and result interpretation can prove difficult, since a low level viraemia can persist for many days. NAAT testing of other clinical samples is possible including sputum, buccal/throat swabs, sweat, urine, breast milk. These samples are not used routinely for diagnostic purposes, except post-mortem when buccal swabs may provide a useful sample. Testing of semen and cervical samples is performed to evaluate the risk of sexual transmission, but usually deferred to the outpatient clinic setting and it is not known whether PCR positivity associates with potential for transmission.

5.7   Supportive Pathology Assays

Laboratory support to monitor for complications of EVD, such as electrolyte disturbance, clotting abnormalities, organ failure is valuable assist clinical teams to correct such abnormalities (see Management). These parameters have been routinely monitored in patients managed in resource-rich settings, either in specialist Category 4 laboratories, or in routine diagnostic laboratories. In resource rich settings, supportive assays have been performed using routine analysers, in standard pathology laboratories, often prior to an Ebola diagnosis being confirmed with no recorded transmissions to laboratory workers. This suggests that risk to laboratory workers processing samples following standard precautions is relatively low. However, care should be taken to follow procedures correctly. In some high resource settings, stand-alone, discrete analysers, such as Point of Care biochemical analysers, are integrated into high-level containment patient care facilities. In low resource settings, bedside, POC tests to exclude malaria and HIV, blood glucose monitoring and monitor electrolytes are a valuable addition to case investigation that do not require laboratory support. As sodium and potassium may vary widely, both above and below the normal range, establishing point-of-care monitoring of these electrolytes to guide supplementation is highly recommended and should be prioritised if resources allow. Urine testing for HCG is also advised routinely.

> **Clinical Scenarios: The Trouble with Testing**
>
> 1. You are clinical officer in a district hospital in rural West Africa. You see a patient in the Emergency Department who is unwell with fever, diarrhoea and vomiting. The patient's relatives tell you that two other people in the village have died in the last week, one of them was vomiting blood. There is no current viral haemorrhagic fever outbreak known in your area. You
>
> (continued)

contact your local public health officer who takes the details and calls you back the next day. He suggests collecting a sample and sending it to the local laboratory, who will arrange testing. You do so, but are concerned the result won't be available for several weeks as it has to go to a foreign laboratory. Meanwhile, five more patients with similar symptoms present to your hospital. You manage to isolate the first two patients, but don't have rooms to isolate the others and so you manage them on the general ward with contact precautions and hope for the best. … Unfortunately, the first patient dies before you have a diagnosis. You try to council the family about safe burial but the family proceed with a traditional burial. The result eventually comes back confirming EVD, you contact the family to notify them, but by now the relatives have left the town on business and you can't locate them. The nursing staff in your hospital are extremely concerned about their possible exposure to symptomatic patients on the general ward.

2. You are a clinical officer in a district hospital during the West African EVD outbreak. You are in charge of a small ward for suspected EVD cases (Holding unit), where patients can be managed in isolation but the ward is full. There is a line of patients outside the hospital waiting to be seen. You screen the patients outside the hospital for symptoms of EVD, but many of them meet the current suspect case definition, including fever, vomiting and abdominal pain, which are similar to malaria and gastroenteritis. You consider admitting them to the General Ward, where there are beds available but you know that might put staff and other patients at risk. You review the list of patients inside your Holding unit. Most of them were admitted yesterday, so you don't expect their results until tomorrow. There is nothing you can do but turn the new patients away and give their families advice on how to manage them at home.

References

1. Baize S, Pannetier D, Oestereich L, et al. Emergence of Zaire Ebola virus disease in Guinea. N Engl J Med. 2014;371:1418–25.
2. Walker NF, Brown CS, Youkee D, et al. Evaluation of a point-of-care blood test for identification of Ebola virus disease at Ebola holding units, Western Area, Sierra Leone, January to February 2015. Euro Surveill. 2015;20:21073.
3. Broadhurst MJ, Kelly JD, Miller A, et al. ReEBOV antigen rapid test kit for point-of-care and laboratory-based testing for Ebola virus disease: a field validation study. Lancet. 2015;386:867–74.

Sequelae of Ebola Virus Disease

6

Patrick Howlett and Marta Lado

Contents

P. Howlett (✉)
King's Sierra Leone Partnership, King's Centre for Global Health, King's College London and King's Health Partners, Freetown, Sierra Leone

M. Lado
King's Sierra Leone Partnership, King's Centre for Global Health, King's College London and King's Health Partners, Freetown, Sierra Leone

King's College London and King's Health Partners, London, UK

© Springer International Publishing AG, part of Springer Nature 2018
M. Lado (ed.), *Ebola Virus Disease*, https://doi.org/10.1007/978-3-319-94854-6_6

6.1   Introduction

The long-term effects of the Ebola virus and the care needed by survivors is an evolving area and in the coming years we have much opportunity to learn about the later complications of the disease. The main concerns in the care of EVD survivors cover a broad range and severity of issues from debilitating and painful physical symptoms, to complex psychosocial issues. At present, there is limited data to confidently provide symptomatic patients with a diagnosis; whether this is a well-known post viral syndrome e.g. post viral fatigue or a syndrome specific to EVD, it is yet to be defined. Even less so are we able to provide evidence based treatments. However, as evidence is gathered and shared, diagnostic confidence will hopefully improve and evidence based treatments will become available. In this chapter, we will define EVD survivors from the time at which they become serum EVD RT-PCR negative and hence, in practical terms, when the patient is discharged from an emergency treatment unit. Although this may not equate to absence of viral replication in the body, it denotes a time at which their acute viraemia has been controlled and their symptoms and focus of treatments will change.

In this chapter, we will first outline some of the proposed pathways for sequalae of EVD, then focus on the body systems in which some of these problems occur and structure these around the time-frame in which they happen. Within the broad range of complications, arguably the most widespread and potentially manageable relate to the psychosocial and ophthalmological sequelae, and viral persistence in semen with the accompanying risk of disease transmission. The following unique aspects of each denotes their importance.

- Psychosocial—Survivors carry a huge mental burden of adjustment disorder, stigma and loss of livelihood, alongside physical symptoms. Often they must face this without the support of many close contacts who may have died of EVD. Psychosocial services should be made available early for all survivors, with a subsequent focus on identifying and supporting those with the greatest burden.

- Ophthalmological—Uveitis is one of the most common complications and can lead to blindness and cataracts however responds well to treatment when identified early. Systems providing ophthalmology follow-up with specific training in ophthalmological complications of EVD should be offered for all survivors.
- Viral persistence in semen—EBoV has been documented in semen up to 18 months following acute disease, with evidence of transmission almost 6 months following acute disease. Semen testing should be readily available for all male survivors and, in the absence of testing negative twice for EVD RT-PCR in the serum, a clear message of barrier contraception should be recommended for at least 12 months.

Given the paucity of data and lack of specific knowledge regarding survivor sequelae, there is not enough evidence to make specific recommendations for subgroups such as concurrently HIV infected, diabetics or indeed pre-adolescent children. However, it stands to reason that we may come to expect differences in specific complications at the extremes of age and as survivors come to develop specific co-morbidities. HIV particular is of concern, given its immune modulating effects and higher prevalence in Sub-Saharan African countries that bear the burden of EVD outbreaks.

6.2   Pathogenesis of Viral Complications

The pathways for complications of acute EVD may be directly virally mediated, or indirectly mediated. Amongst the directly virally mediated complications, symptoms may correspond to ongoing viral persistence or dysfunctional, ongoing immune response. To consider the interaction between persistence and the immune response, we must first outline documented viral persistence and current evidence of the immune response in EVD survivors.

6.2.1   Viral Persistence

Evidence of viral persistence in bodily fluids has been clinically well documented, the longest period of EBoV RT-PCR positive results in semen is 18 months following disease [1] and viability proven through transmission at least 179 days after symptom onset [2]. Factors affecting latency of viral persistence remain unclear, with the case of EBOV RT-PCR semen positivity at 18 months being in a man with well controlled HIV [1]. The source of ongoing viral replication in the genital tract is unknown however may be the testes, prostate or accessory gland. With time from initial discharge, the viral load detected and proportion of survivors with positive samples decreases. There appears to be no effect of antiviral therapy on length of persistence. Models suggest that 50% percent of men will clear EBoV RNA from their serum by 115 days and 90% by 295 days. Correlated to the West African outbreak, this will translate to a single male with positive EBoV RNA in the semen by July 2016 [3].

The current WHO guidance on semen testing is presented in Box 6.1. Trials are currently under way to use the novel nucleotide analogue prodrug, GS-5734, to clear virus from the semen of persistent EBoV RNA in the semen.

Box 6.1 Current WHO survivor Recommendations: Guidelines for Semen Testing and Counselling (Adapted) [4]

- The semen of EVD survivors should be assumed to contain Ebola virus for the first 3 months after disease onset.
- RT-PCR testing of the semen should then be performed at 3 months and every month thereafter until their semen tests negative for virus twice, with an interval of at least 1 week between tests.
- Pre-and post-test counselling should be performed by experts in sexual transmission counselling.
- Until a male EVD survivor's semen can be determined to be Ebola virus free through the testing described above, survivors and their sexual partners should either (a) abstain from all types of sex or (b) observe safe sex through correct and consistent condom use.
- Having tested twice negative, survivors can safely resume normal sexual practices, although condom use is still recommended.
- If an Ebola survivor's semen has not been tested, he should continue to practice safe sex for at least 12 months after the onset of symptoms.

EVD virus has also been identified through RT-PCR in a variety of body compartments including the saliva (22 days), conjunctiva/tears (28 days), stool (29 days), vaginal fluid (33 days), sweat (44 days), urine (64 days), amniotic fluid (38 days), aqueous humor (101 days), cerebrospinal fluid (9 months) and breast milk (16 months) [5] (see Fig. 6.1). Interestingly however, despite records of several samples being sent, synovial fluid has not yet been identified as a reservoir [6, 7].

In addition to inferred viability through sexual transmission, positive viral culture with symptomatic disease, following clearance from the serum, has been found in the cerebrospinal fluid at 9 months and the eye 9 weeks following symptom onset. It should be noted that lack of culture viability does not denote an absence of live virus and hence lack of transmissibility. However, in cases where sweat and urine have tested RT-PCR positive, patients were discharged with standard precautions provided to the patient and family and no subsequent cases in contacts were reported [6, 8].

6.2.2 Immune Response

EBoV infection is linked to a pronounced and sustained humoral and adaptive immune response. During the initial disease and viraemia, the IgM and IgG responses develop early with IgM detectable as early as day 2 post symptom onset

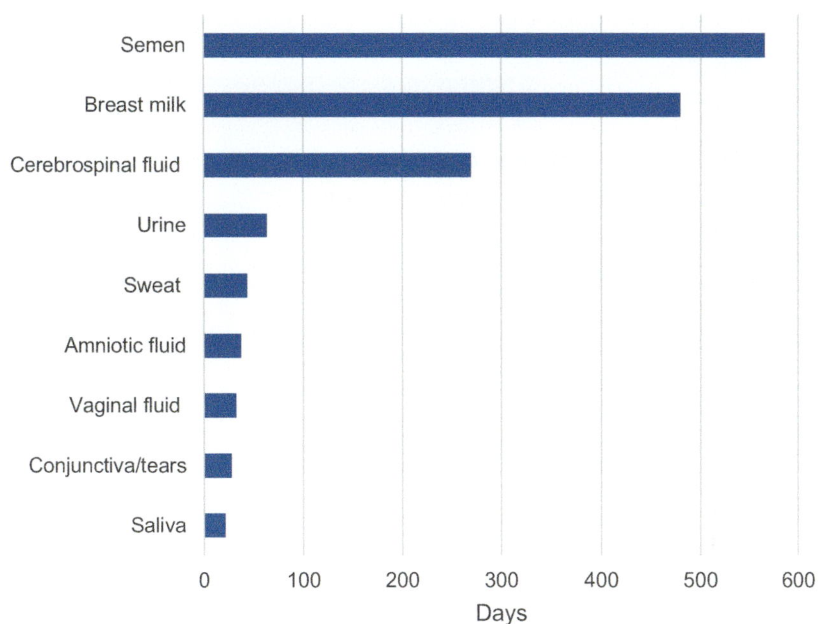

Fig. 6.1 Virus persistence in days after disease onset, as measured in body compartments of survivors of Ebola virus disease, detected by RT-PCR

and persisting up to 24 weeks, and IgG as early as day 6 before persisting for up to 12 years [9, 10] (see Fig. 6.2). Activated CD4 T cells and, more notably, CD8 T cells proliferate early during acute disease. In more severe cases of disease, activated CD 8 T cells persist up to 60 days after symptom onset—far longer than expected when compared to other virus infections [12].

Levels of IgG decline at differing rates. At 29 months following EVD, higher levels were seen in those with arthralgia while a complete absence was seen in 22% of survivors [10]. In addition, ring vaccination studies have shown correlation between Ebola specific IgM and IgG antibody levels and EVD protection up to 21 days following vaccination [13]—the extent to which this antibody response is protective of acute disease, especially in comparison to those previously infected with EBoV will be of interest.

6.2.3   Survivor Symptoms, Viral Persistence and the Immune Response

Early evidence suggests symptoms experienced by EVD survivors are associated with acute viral load, clinical site of EVD disease and a persistent inflammatory response. Patients with higher viral loads were more likely to experience eye symptoms [14]. The presence of the same symptoms during acute disease and subsequently afterwards as a survivor has also been reported [15]. As noted in the

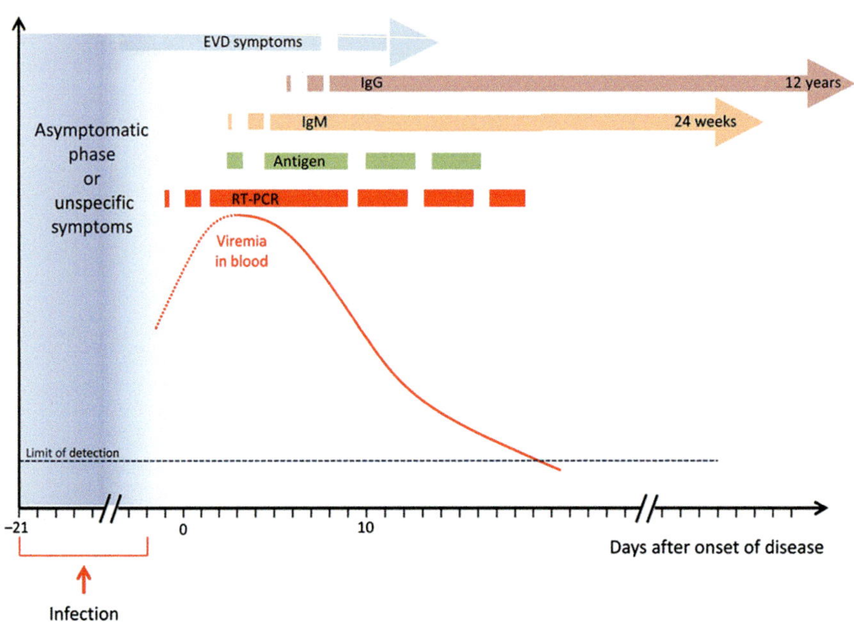

Fig. 6.2 Evolution of viral load in blood and appearance of immune response during the course of Ebola virus disease (EVD). *IgG* immunoglobulin G, *IgM* immunoglobulin M, *RT-PCR* reverse transcription–polymerase chain reaction. (Note to editor: Please redraft this image—do not have authorization to copy from [11])

previous section, activated CD8+ T cells were found 60 days after symptom onset in a patient with severe disease, in comparison patients with mild disease had rapid and early contractions of T cell populations [12]. This persistent activated adaptive response is particularly suggestive of ongoing antigenic stimulation to the presence of "reservoir sites" of live virus, or viral fragments outside of the serum. Clinical observations would suggest a significant inflammatory component to certain survivor sequelae, such as arthralgia or uveitis, with response to anti-inflammatory treatment [4]. We have also seen that there is persistence of virus or virus particles up to 18 months [1] and that there is an activated immune response up to 29 months [10].

How the virus and immune response interacts to produce the breadth and severity of symptoms of the "Post-Ebola Syndrome" remains unclear, however may relate to one of more of several aetiologies.

Firstly, in a similar way to rubella or enterovirus infection, the activated response and symptoms are mediated by active viral replication. This hypothesis is supported by evidence of replicating and live EBoV virus with accompanying symptoms and transmissibility in the eye, CNS and semen of EVD survivors, and rhesus monkey models [16].

Alternatively, as is seen in alphaviruses (including Chikungunya, in which the post viraemia arthralgia may be particularly prominent and severe) and parvovirus infection, an immune complex deposition of a whole or part of the virion is deposited in joints or tissues as part of initial humoral immune response, and leads to persistent immune activation and symptoms. In contrast to active viral replication, samples will be viral culture negative—as has so far been demonstrated in EVD survivors with arthralgia and synovitis who have had synovial specimens sent for RT-PCR testing.

A final option may relate to the intense and invasive EBoV disease with the accompanying profound inflammatory response. This may leave post infectious damage, such as deafness (as suggested in Lassa Heamorrhagic fever) or, potentially, cerebral atrophy.

6.2.4   Indirectly Mediated

In addition to virally mediated effects, EVD carries many of the same physical and neuropsychiatric sequelae of other acute, infective critical illness, for example thrombophilia, critical care neuropathy and post ITU adjustment disorder. In the absence of data comparing critical care patients from similar settings, being able to delineate generic critical care conditions from those specific to EVD may prove challenging.

Within the context of the most devastating West African outbreak ever seen, and in a setting where resources were limited, the ability to identify potential iatrogenic complications was limited. However, these may become more apparent with time and relate to novel medications or the extensive use of high concentrations of chlorine, as reported in the case of chlorine inhalation lung injury in a Ebola healthcare worker [17].

6.3   EVD Survivor Group Characteristics

Although patients aged 12 to <18 years experienced the highest survival rate (61%), the median age of survivors is 28 years (see Fig. 6.3 Source Liberia MOHS, n = 1536) with an approximately equal split of male and female (Male 46%, Female 54%). In an already young population and as those aged ≥65 years had the lowest survival (17%) [18], this latter group is least well represented [19]. HIV infection, at the time of discharge, seems rare with few reported cases of HIV positive survivors [20].

6.4   EVD Sequelae

This relatively young and previously relatively healthy group may present with a broad range of sequelae, which can be considered according to the likely time of presentation (Table 6.1), and system of disease.

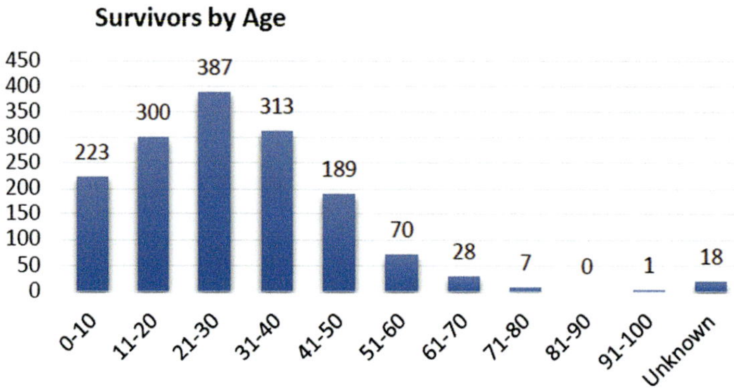

Fig. 6.3 Demographics of EVD survivors in Liberia. Source Liberia MOHS, n = 1536

6.5   Multisystem

6.5.1   Sepsis, Malaria Co-infection and Respiratory Failure

Patients cared for in high income settings have been shown to develop gram-negative septicaemia in the immediate peri-viraemic stage. The source of the bacteremia is thought likely from gut translocation, supported by ultrasound findings of an oedematous bowel wall and associated paralytic ileus. In the acute setting, this has led to recommendations for empirical treatment with broad spectrum antibiotics during the acute disease [21].

In sub-Saharan African settings, co-infection and hospital acquired infection with malaria is common with estimates of co-infection during the acute stage of EVD ranging from 4.4 to 19% [18, 22]. Although an initial study found improved outcomes in those with malaria co-infection, this has largely been refuted by subsequent data and mortality is likely higher in malaria co-infection [18, 23, 24]. In the post-viraemic setting a survival benefit is similarly less likely and vigilance is required, especially where blood products are being administered.

In the periviraemic stage, factors such as pulmonary atelectasis, volume overload (more likely in high income settings), and aspiration, leading to acute respiratory failure, have been managed by invasive and non-invasive ventilation [8]. Further respiratory compromise by aspiration of blood from epistaxis in the context of thrombocytopenia has also been reported [25]. These complex and multisystem issues would suggest that patients in the immediate post serum EBoV positive phase should be managed in a high dependency setting, where available.

6.5.2   Malnutrition, Weight Loss and Anorexia

Due to a multitude of factors including an extreme catabolic state, decreased consciousness, and lack of access to invasive feeding and hospitalization, patients commonly

Table 6.1 Reported post EVD sequelae, presented according to most common time of presentation, and disease system

	Acute (0–8 weeks)	Late (>8 weeks)
Multisystem	Bacterial infection Malaria (4–19%) Malnutrition and weight loss (10%) Anorexia (7–25%) Fatigue Thrombophilia Disability Death (3%)	Fatigue (29%) Disability
Mental health and psychosocial support	Stigma Adjustment disorder Depression/Anxiety	Stigma Depression (21%)/Anxiety
Ophthalmology	Uveitis (11–62%) Cataracts	Uveitis (11–62%) Cataracts (10% of those with uveitis) Retinal scarring (15%) Visual loss (0.2–3%)
Sexual and reproductive health	Miscarriage and Stillbirth (90%) Decreased libido (23%) Viral persistence in semen Vaginal candidiasis (20%)	Miscarriage and Stillbirth Viral persistence in semen Vaginal candidiasis (10%)
Rheumatology	Arthralgia (76–87%) Arthritis/Synovitis Myalgia (14–38%)	Arthralgia (27–76%) Myalgia (14–38%)
Neurology	Headache (35–88%) Stroke (0–2%) Encephalopathy (<1%) Meningo-encephalitis (<1%) Seizure	Headache (35–88%) Meningo-encephalitis (<1%) Peripheral parastesia/dysthesia Hearing loss (2–7%)
Dermatology	Desquamating rash (27–33%) Alopecia (1%)	Desquamating rash (27–33%)
Gastrointestinal	Abdominal pain (32–35%)	Abdominal pain (12–35%)
Cardiology	Tachycardia Pericarditis/Myocarditis (<1%)	
Respiratory	Respiratory Failure	

Reported prevalence according to clinical features are presented in brackets alongside. Where no estimate is given, insufficient evidence exists to make an estimate

have a striking weight loss on discharge with over 10% having lost over 5 kg during their short admission [26]. Compounding this, survivors present with almost universal anorexia following discharge, the majority considering this moderate to severe [27]. With time this improves and the majority regain weight in the following weeks to months, although in a minority of cases this persists up to 6 months [10]. Models of good care in low resource settings include blended, high calorie foods for 2 weeks following discharge, followed by cash supplements for food for the following 2 months.

6.5.3   Fatigue

Pooled analysis suggests 29% of survivors may be affected by fatigue [28]. It may feature at any time from discharge and is one the few symptoms shown to be significant at 2 years when compared to household contacts [29]. Fatigue does appear to improve with time without any specific treatment (Healy—personal communication). No assessments have been made yet to link symptoms to post-viral fatigue, or indeed attempt any interventions.

6.5.4   Disability

Disability, across multiple domains most significantly in mobility, vision and affect has been found to be more prevalent in survivors when compared to controls [30]. However, the extent to which this impacts upon daily function is less clear [31]. Certainly, there is a group who are affected by their symptoms with high levels of disability. This group often has conditions such as stroke, neurocognitive impairment or arthritis, and co-exist with mental health conditions such as depression and anxiety [32]. Improvements have been reported in EVD stroke and neuropathy patients following referral to physiotherapy services.

6.5.5   Thrombophilia

Although cases are limited, several reports exist of stroke or pulmonary embolus/deep vein thrombosis either during admission or in the few days following discharge [7, 33]. This is supported by thromboelastography (TEG) studies, performed on two patients during the viraemic and periviraemic stages of disease [34]. Although during the acute viraemia a consumptive coagulopathy was observed and prophylactic low molecular weight heparin was withheld, in the immediate stages following this both patients developed a hypercoagulable state, in one case with a marked thrombocytosis (1726×10^9—day 21 illness). Both patients were treated with enoxaparin, as heparin resistance was observed on TEG testing for 2 weeks post discharge. The thrombocytosis was treated with low dose aspirin 75 mg OD.

6.5.6   Death

One study that followed up 151 survivors found 4 deaths within the 6 weeks immediately following discharge, these were attributed to TB, stroke, pancreatitis and chest infection [35]. Although this data is not case controlled, this relatively high 6 weeks mortality rate suggests particular care should be taken in following up EVD survivors for all causes of morbidity and mortality in the first 6 weeks.

6.6   Mental Health and Psychosocial Support

Probably the most important aspect of survivor care and support centers around the psychosocial and mental health assessment and support. Although survivors are generally thankful and happy to have survived, there are, nevertheless, unintended and negative psychosocial consequences. Understanding the complexities of these issues and the relevant response requires multidisciplinary involvement and intervention (see Fig. 6.4). Furthermore, the co-ordination and implementation of a response is best approached through the involvement of communities, government, NGO's, and, most importantly, the survivors themselves.

For those survivors, the majority will have been through exceptionally harrowing experiences and have lost many friends and family. When discharged, they are faced with stigma from their family and community. This, combined with a loss of livelihoods, financial pressures and a new constellation of physical symptoms, means the adjustment to a new life as an Ebola survivor is an incredible challenge which can lead to stress and predispose to mental illness. Almost all report that, on

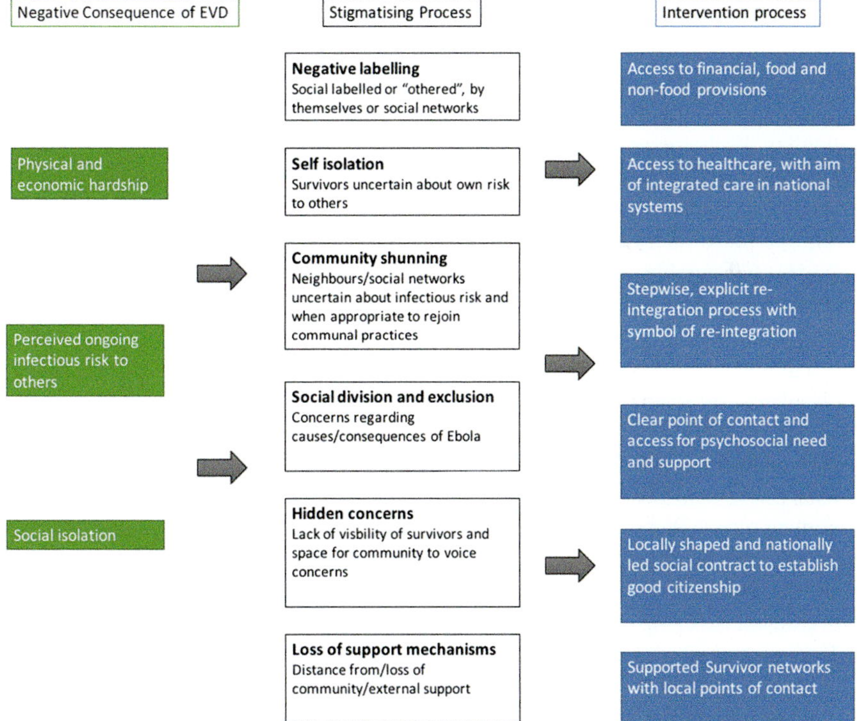

Fig. 6.4 Possible negative consequences, stigmatizing processes and intervention processes for alleviating stigma for EVD survivors (adapted with permission from Ebola Anthropology Platform. Ebola Survivors using a stepwise re-integration process to establish social contracts between survivors and their home communities)

discharge, they feel less confident within themselves and feel that their disease has had a negative impact on their social life [27].

6.6.1 Stigma

Stigma experiences reported survivors include; rejection by friends and family with, in some cases, divorce; being prevented from attending their places of worship; and people not wanting to buy their goods or touch their money [27, 36, 37]. For those in the community, especially in the early stages following an epidemic, there is a widespread fear of infection from survivors and many people in the community report being unwilling to shake hands or hug an Ebola survivor [38, 39]. These experiences lead to feelings of rejection, humiliation and isolation that persists in a large proportion well up to a year after discharge [15, 40].

6.6.2 Depression

Pooled rates of depression amongst survivors are 21% [28], however background rates are not readily available for comparison. Survivors may present 1–24 months following discharge and severe cases of depression are characterized by low mood, feelings of low self-worth, sleep difficulties, hallucinations and suicidal ideation or attempted suicide. Those with physical complaints that affect activities of daily living are more likely to present with depression and generalized anxiety disorder [32, 41]. Favourable responses have been seen to standard pharmacological and cognitive behavioural therapies, with preferred follow-up being through local mental health networks.

6.6.3 Lessons Learnt from the West African Ebola Outbreak

The most recent West African Ebola outbreak presented an unprecedented challenge of re-integration of survivors and providing psychosocial support, and reflections on what did and did not work are still being learnt. Higher profile response include Government led campaigns to reduce stigma, such as those led by the President Ernest B Koroma and Sierra Leone State House (Fig. 6.5) and toolkits provided by the WHO for Psychological First Aid [42] (see Fig. 6.6) and USAID for Stigma Reduction [43]. Longer-term however, the majority of the focus must be on support of and integration into existing health systems support to ensure that psychosocial and mental health care is available for survivors and non-survivors.

Fig. 6.5 Image of Sierra Leone State House Advert for Survivor reintegration (courtesy of Marta Lado)

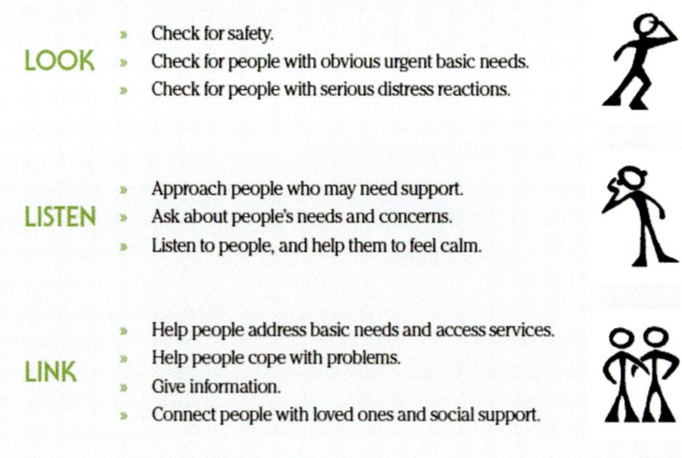

Fig. 6.6 Principles of Psychological First Aid, adapted from WHO Toolkit

6.7   Ophthalmology

6.7.1   Uveitis

Between 11 and 62% of survivors have been reported to develop uveitis [7, 14, 29, 44]. The median time to onset is 2–3 weeks post discharge, although symptoms may present during acute disease and up to 4 months post-discharge. Indeed, in longer-term studies, 39% of survivors still report blurred vision up to 29 months post-discharge.

Uveitis is most commonly anterior however 25% may be posterior and a further 25% may be panuveitis. Uveitis in survivors has been found to be more common in patients with red or injected sclera, fever, and lower cycle threshold values during acute admission. On screening, asking about blurry vision, light sensitivity, or itchy eye(s) was found to be 88% sensitive and 50.7% specific in EVD survivors [14]. Common differentials include bacterial, viral or allergic conjunctivitis and trauma. Care should be taken to exclude herpes simplex keratitis as this worsens with steroid treatment. Ideally, this is achieved with a fluorescein stain showing a classical linear branching ulcer. Uveitis is immediately sight-threatening and treatment is time sensitive. Furthermore, delays and lack of treatment increase the risk of cataracts.

Treatment is with topical steroids, such as prednisolone, and mydriatics, such as cyclopentolate or atropine. Cases of anterior uvietis that do not improve in 5–7 days should be prescribed oral steroids and a specialist referral should be made. Cases of intermediate, posterior or panuveitis should be managed more intensively with a view to early specialist referral. In one case of anterior uveitis presenting 14 weeks after symptom onset that was not responsive to topical treatment, a vitreous paracentesis was perform and the fluid found culture positive for EBoV. The initial cycle threshold value was 18.7 and a conjunctival swab, taken at the time of paracentesis from the vitreous humour was negative for EBoV. The patient received combined oral and topical steroid treatment and oral Favipravir, an experimental anti-RNA drug, and symptoms resolved [45].

Box 6.2 Symptoms, Examination Findings and Management of Uveitis in EVD Survivors (Adapted from WHO Survivor Care Guidance)

		Intermediate Uveitis	Posterior Uveitis	Panuveitis
Symptoms	Anterior Uveitis			
Symptoms	Pain, redness, blurred vision and increased lacrimation	Minimal redness, itchiness with floaters and blurred vision	Absence of pain and redness. Floaters and blurred vision present	Features of both anterior and posterior uveitis

(continued)

Box 6.2 (continued)

Inspection findings	Perilimbal injection, decreased visual acuity, pupillary miosis. Increased ocular pressure (see Fig. 6.7)	May have some perilimbal injection	Normal on inspection	Signs of anterior uveitis, particularly perilimbal injection
Slit-lamp examination	Keratic precipitates, Corneal edema and vitreous haze	Inflammatory exudate on posterior wall (pars plana)	Inflammatory exudate on posterior wall (pars plana)	Findings of both anterior and posterior uveitis
Treatment [4]	Prednisolone 1% 1–2 hourly Cytopentolate 1% 4 times/day. If no improvement in 5–7 days, convert to oral steroids and refer to specialist	Prednisolone 1% 1–2 hourly Cytopentolate 1% 4 times/day Consider early conversion to oral steroids and referral to specialist care		

Early cataracts presenting within 2 weeks of discharge appear a common complication of EVD survivors who have symptoms of uveitis, with up to 10% of those with uveitis having early signs, despite a median age of 29 years [14]. The cataracts are associated with low intra-ocular pressures in 80% of patients, suggestive of impaired aqueous humour production [46]. This finding casts uncertainty over the symptomatic benefit of cataract surgery in the low pressure group.

Concerns exist regarding the risk of infection during surgery and pathways are yet to be put in place to treat this growing number of patients. The limited evidence that does exist suggests that the EBoV virus is not persistent in the aqueous humour of

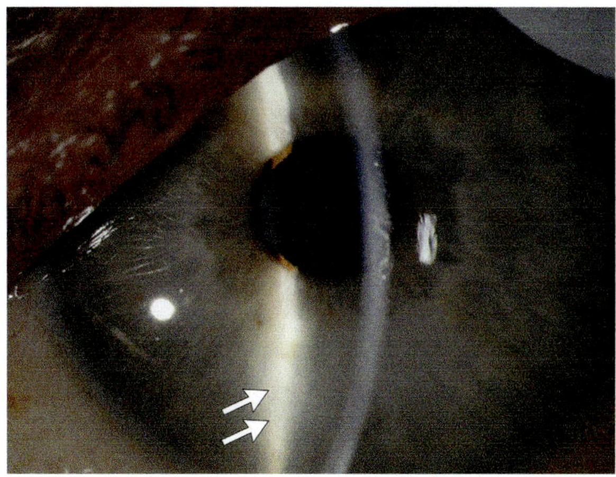

Fig. 6.7 Slit-Lamp Photograph of the Left Eye 14 Weeks after the Onset of Ebola Virus Disease: Mild corneal edema, rare keratic precipitates (arrows), and inflammatory cells and protein in the anterior chamber— From Varkey JB et al. N Engl J Med 2015;372:2423-2427. Copyright © (2017) Massachusetts Medical Society. Reprinted with permission from Massachusetts Medical Society

these cataract patients [46], although current opinion would still suggest sampling of the aqueous humour prior to surgery.

6.7.2 Other Ophthalmological Complications

Isolated symptoms of blurred vision, photophobia and retro-orbital pain are common and may still be present years after the acute phase of disease. Although following the Sudan Ebola Virus outbreak in 2000 in Gulu, Uganda 60/257 (23%) survivors reported visual loss, this has not been repeated in the current outbreak where 0.2–3% have so far been found to have complete visual loss [14, 15, 27]. The causes are likely multifactorial, including uveitis, cataract and, in rare cases, cortical blindness. The most recent epidemic has also uncovered the potential for retinal changes. In one case-control study of patients presenting with ocular symptoms, a well-demarcated, wedge or diamond shaped dark lesion was present in 15% of survivors, but none of the control. In cases where the lesion was located adjacent to the optic disc, the curvilinear projection appeared to align with the pathway of the retinal ganglionic axons, suggesting a neurotrophic spread through the eye [46].

6.8 Sexual and Reproductive Health

Outcomes for women and the unborn fetus in acute EVD are not systematically described but likely to be poor with approximately 90% mortality for the mother and nearing 100% for the newborn [4]. Only one live birth during the acute phase of disease has been reported in a mother treated with experimental therapy [4]. Further details on acute EVD in pregnancy are presented in Chap. 5.

6.8.1 Women Infected with EVD During Pregnancy

As a result of the poor outcomes for pregnant mothers with acute EVD, few reports exist of pregnancy in the post viraemic period. Those available describe stillbirths, with day of delivery ranging from between 7 and 32 days post discharge from ETU, and in mothers infected at between 5 and 7 months gestation. Deliveries were associated with positive EVD RT-PCR from the stillborn fetus, amniotic fluid (although 1 tested negative), meconium and cord blood, all with strong positive Ct values. Two cases were delivered in an Ebola Treatment Unit, with full EVD PPE, another delivered at a local health centre by a traditional birth attendant using gloves and an apron. No secondary infections were reported from these cases [47, 48].

Nevertheless, given the risk of stillbirth occurring at any time in the subsequent gestation period and potential secondary infections, WHO recommendations for the care of women infected with EVD during their pregnancy [4] are:

- Provide close clinical follow-up and monitoring, with significant emphasis placed upon counselling regarding risks to mother and fetus.

- Immediate arrangements should be made for safe transfer to a setting where full EVD PPE can be used, specimens and waste can be safely managed, and complications of labour can be managed.
- If chorioamnionitis is suspected, proceed with labour induction regardless of gestational age
- Cord blood and swabs of the products of conception (neonate, placenta and amniotic fluid) should be tested for EBoV by RT-PCR.
- In the case of survival of a newborne and as symptoms of EVD in neonates may be atypical or not evident early in infection, the infant should be managed using standard EVD IPC precautions for 21 days, regardless of laboratory results or the presence or absence of symptoms.

The management of Pregnant women infected with Ebola is covered in Chap. 5. During labour, the principles of management for women with acute infection and those who were infected during their pregnancy but deliver afterwards should be the same.

6.8.2   Women Who Conceive Following Recovery from Acute EVD

One short report followed 70 female survivors who became pregnant within 15 months of discharged after acute disease. Of these women, 49 pregnancies resulted in normal births, 15 miscarried, 4 were stillborn and 2 chose to terminate their pregnancies. Of note, all stillbirths occurred pregnancies conceived within 6 months of discharge [49]. Cases were managed in standard healthcare facilities and no further infections were reported. No evidence exists for risk factor during acute disease which may influence subsequent maternal outcomes.

WHO recommendations are that pregnancies should be cared for according to WHO and national recommendations and standard obstetric IPC precautions should be followed, however given the height rates neonatal complications (19/68, 28%), it may be prudent to follow up this group more closely, especially if conception has occurred within 6 months of discharge.

6.8.3   Breast Feeding

Low levels of viral persistence in breast-milk has been reported up to 16 months following acute disease, although infectivity has not been shown. WHO recommendations currently suggest that women should have the option to test breast milk for EBoV RNA. Should they test positive, the breast milk should be retested every 48 h until two negative tests are obtained, whereupon breast feeding can resume. Babies who have breastfed from a mother whose milk tests positive should be monitored as a close contact for 21 days since their last day of breastfeeding from the EBoV positive milk. Those who are unable to choose to test or choose not to test can continue breastfeeding.

6.8.4   Sexual Health

In the early stages following discharge, vaginal candida responding to topical clotrimazole occurs in 20% of female survivors. Incidence decreased to 10% in the following 10 months (Healy—personal communication). Amongst men, loss of libido is common affecting almost ¼ in the early stages following discharge, however this does not appear permanent [27].

6.9   Rheumatology

6.9.1   Arthralgia, Arthritis and Synovitis

Along with ophthalmological complaints and headache, arthralgia is the most frequent physical symptom noted by EVD survivors. Between 50 and 75% of survivors describe joint pains, with onset starting from as early as immediately following their acute admission, and persisting up at least 29 months in up to a third of patients [7, 10, 14, 29, 40]. The majority affected describe an asymmetrical, large joint polyarthralgia which may come and go over time, and may or may not improve during the day. In a small number of reviewed survivors, the joints affected—in order of frequency—were knees, back, hips, fingers, wrists, neck, shoulders, ankles, and elbows [10]. Associated features in a small number may include evidence of joint inflammation including pain, redness and effusion. Some have suggested the concurrent findings of arthralgia, periarticular tendonitis (enthesitis)—often affecting the shoulders and hips (sacro-ilietis)—and uveitis provides an overall picture consistent with a spondyloarthritis.

So far, any plain radiographs of the affected joints have failed to show any significant joint changes and the only reported synovial tap did not shown evidence of EBOV RNA [6]. Inflammatory markers are almost always normal, as has been described in other post-viral arthralgias [4].

The principle aims of managing arthralgia in EVD survivors are; the exclusion of alternative causes; analgesia; anti-inflammatory treatment (with consideration of drug side effects); physical rehabilitation; review in the context of ongoing psycho-social issues; and consideration for specialist review. It should be noted that the majority of Sub-Saharan African countries have a high prevalence of H. Pylori and unrestricted access to non-steroidal anti-inflammatory drugs, as such care should be taken in prescribing and educating patient on the use of potentially gastro-erosive medications to minimize the risk of ulceration.

Box 6.3 Management of Arthralgia in EVD Survivors (Adapted from WHO [4])

- **Exclusion of alternative causes**—history, physical examination and investigations should consider common infectious presentations such as septic joint, malaria and HIV, and non-infectious such as Rhuematoid Arthritis, Systemic Lupus Erythematosus or gout.
- **Analgesia**—non-pharmacological methods include a cold/warm compress. First line medication for adults and children is Paracetamol (max 1 g three times per day for adults, 15 mg/kg three times per day for children—note reducing dosing schedule due to concerns of sub-maximal liver function, especially in early stages post acute viraemia).
- **Anti-inflammatory medication**
  - Second line analgesia for adults is Ibuprofen 200–400 mg orally up to three times daily. Other acceptable NSAID regimens include diclofenac 50 mg orally two or three times daily or naproxen 250–500 mg orally twice daily. Indomethacin is not recommended due to greater risks of gastric complications. Children should be prescribed Ibuprofen 10 mg/kg orally up to three times daily. Considerations to take into account regarding treatment include the risks of gastric ulceration (and need for prescription of a proton-pump inhibitor), dose adjustment in renal impairment and monitoring of renal function.
  - If significant or disabling symptoms persist after 7–10 days adults may be prescribed 20 mg oral prednisolone for 7–10 days. For children, referral should be made to the nearest specialist survivor centre. Consideration to take into account regarding treatment include risks of hyperglycaemia, gastric ulceration, mood changes and weight gain. And co-existing strongyloides infection, HIV and TB infection and pregnancy status. Review at 1 week following treatment for improvement in symptoms and potential side-effects is recommended.
- **Physical Rehabilitation**—refer early to local physiotherapy services, and recommendation to mobilise as safely tolerated.
- **Review in the context of ongoing psychosocial issues**—physical symptoms have been shown to overlap with low mood and a holistic approach to management should include sensitive questioning regarding grief, stigma, current work status, and anxiety and depressive symptoms.
- **Indications for specialist review**
  - Arthralgia that does not respond to NSAIDs and/or prednisolone treatment
  - Spinal and sacroiliac joint involvement
  - Requirement for intravenous antibiotics, or need for joint aspiration. Note, joint aspiration should be performed using EVD precautions and aspirate sent for RT-PCR testing for EVD prior to further testing
  - Cases of diagnostic uncertainty or prolonger musculoskeletal pain or fatigue

Although joint pains seem prevalent amongst survivors and at 6 months almost two-thirds of survivors have described the joint pains as a major health problem [10], outcomes seem reassuring. Good responses are seen to first and second line therapy and physiotherapy, minimal long-term complications are reported, and the majority and able to continue with activities of daily living (Healy—personal communication).

6.10 Neurology

Central and peripheral nervous system complications have been observed in EVD survivors, with presentation at both early and late stages. Importantly, these have been shown to be associated with significant disability and in rare cases affect mortality outcomes in a small subset of patients [32].

6.10.1 Delayed Neurological Recovery/Encephalopathy

Although the majority of patients recover quickly, a small number show features of encephalopathy in the first 1–3 weeks following resolution of viraemia. If should be noted however that delineating primary central nervous system disease caused by EBoV and other metabolic or multisystem complications of EVD has proved difficult. The encephalopathic stage is characterized by slowed mental responses, neurocognitive impairment, and delirium with hallucinations, and may takes days to several weeks to fully resolve. Although the majority of cases seem to return to a good level of function, with limited disability [31] and recovery of cognitive function [50], at least one isolated case exists of permanent cognitive impairment, blindness and deafness [32]. Management of this transient neurological impairment is supportive. The associated delirium has, in one case, been found to be resistant to haloperidol although its effective use was widespread amongst those with acute disease suggesting it should still be considered [8].

6.10.2 Meningo-encephalitis

Menigo-encephalitis during acute disease, and early and late recovery is recognized. Presenting features are in keeping with any meningo-encephalitis, including headache, fever, neck stiffness, with subsequent development of fluctuating levels of consciousness and agitation. An additional feature, which may raise the index of suspicion is uveitis. Examination features include decreased conscious level, increased tone and hyper-reflexia, frontal lobe release signs and cerebellar features. Seizures have also been reported, controlled with standard anti-convulsant therapy [32, 51, 52]. In the case of relapse in the CSF, isolated cranial nerve palsies were also noted and interpreted as evidence of skull-base leptomeningeal inflammation—later confirmed on MRI scanning [52].

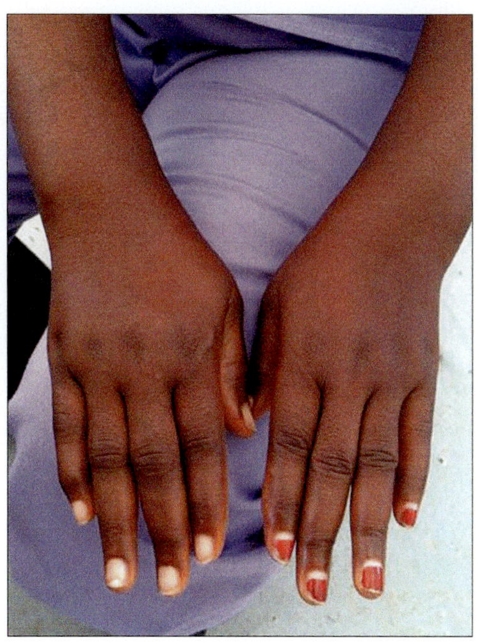

Fig. 6.8 Swollen hands in a 41-year-old woman 7 months after acute Ebola virus disease. The patient reported severe symmetrical pain in the hands, feet, and knees, and markedly decreased vision in one eye. She noted occasional mild joint pain without swelling or tenderness before having Ebola virus disease. The joint pain and diminished vision improved with oral corticosteroid treatment. Photograph by Cornelia Staehelin. Reprinted from The Lancet Infectious Diseases Volume 16, Issue 6, Pages e82-e91 (June 2016) DOI: 10.1016/S1473-3099(16) 00077-3 with permission from Elsevier

Lumbar puncture data is minimal however, opening pressures appear raised between 25 and 38 cm H_2O and RT-PCR is positive for EBoV. In the single reported case of EVD relapse presenting with meningitis symptoms, serum was also RT-PCR positive for EBoV. Computed Tomography imaging in patient with CNS disease suggest some association with cerebral atrophy [32, 53] (see Fig. 6.8). MRI at the time of acute CNS relapse shows evidence of lepto-meningeal enhancement [52]. MRI performed following resolution of acute CNS disease has shown multiple punctate lesions in the cerebral white matter, suggestive of microvascular occlusion and ischaemia [51]. A case study of meningoencephalitis is presented in Box 6.4.

> **Box 6.4 Case Study:**
> A 35-year-old female had a 22-day acute admission for Ebola virus disease, during which she initially improved before becoming unconscious. At this point, days 28 and 29 after symptom onset, serum tested negative twice for EBOV PCR and she was discharged from her ETU. Following slow improvement in her conscious level, on day 41 a lumbar puncture was performed with opening pressure 30 cm H_2O. Her CSF tested RT-PCR EBOV positive (C_t value 37.6) She was diagnosed with a late onset EBOV Encephalitis, which was managed conservatively. Concurrently, she developed an asymmetrical large joint polyarthritis affecting her right shoulder, left elbow, and left knee. A joint effusion from her left knee was aspirated and was negative for EBoV

(continued)

Box 6.4 (continued)

RT-PCR. She was given diclofenac (50 mg) twice daily and, following minimal improvement one intramuscular dose of methylprednisolone (80 mg) with gastro-protection. On day 51, a midstream urine sample was EBOV RT-PCR–positive (C_t value 35.7), and an underarm sweat swab sample was EBOV RT-PCR–negative. The patient was discharged; her family was advised to minimize contact with her body fluids. In the specialist neurology clinic, 435 days after initial diagnosis, she presented with a new all-over headache, worse in sunlight and on carrying loads. Her MMSE, 18/23 on discharge from her acute admission, had improved to 26/27. Her WHO-DAS 2.0 score was 12.5, limited by headaches. Her CT scan (see Fig. 6.9) shows disproportionate cerebral and cerebellar volume loss. Her retinal scan was unremarkable. She was diagnosed with common migraine and noted improvement in symptoms with propranolol 20 mg once daily.

This case provides examples of; late onset meningo-encephalitis with encephalopathy with supportive management, arthritis and synovitis managed with anti-inflammatory medication and steroids, viral persistence in the urine and sweat, headache managed with analgesia and b-blockers.

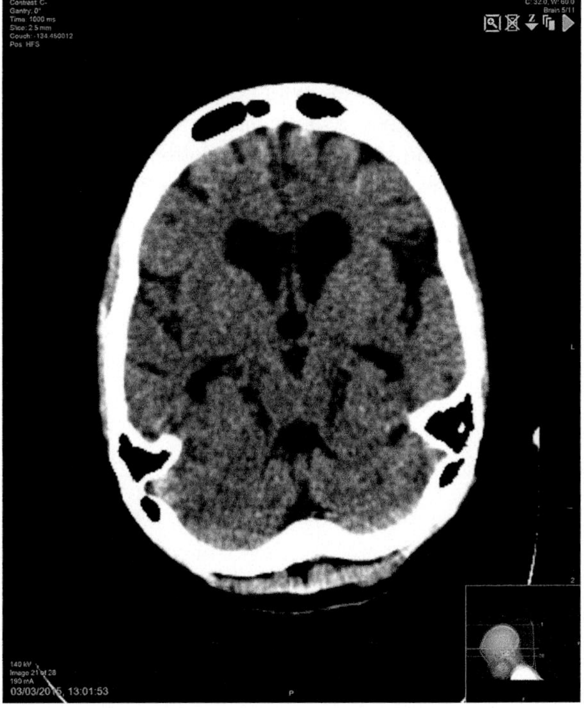

Fig. 6.9 Non-contrast brain CT scan. Female survivor, aged 35. Global (cortical and subcortical) atrophy in keeping with nonobstructive ventriculomegaly and no periventricular low attenuation

Meningo-encephalitis at any stage of EVD disease is potentially life-threatening and has been associated with disability. In addition, there is a potential risk that the patient may be infective, as a viral CNS infection "spills-over" from the immune-privileged site of the CNS into the serum. However, no cases have been reported from survivors managed for EVD CNS infection. As a result, it is advisable to treat EVD survivors presenting with fever and clinical signs of meningitis and/or encephalitis in an isolation facility, until serum testing has been completed. The role for specific treatments is unclear. Two cases have reported the use of steroids (IV Dexamethasone and IM methylprednisolone), one in conjunction with a novel nucleoside analogue antiviral drug [6, 52]. Both cases reported improvements. Until further studies have been done however, no firm recommendations for specific treatments can be made.

6.10.3 Headache

The most prevalent neurological sequelae is headache, which can present early following discharge from isolation, however more commonly presents as a later complication, several weeks or months following the acute episode in between 35 and 88% of survivors [28]. No one characteristic headache exists, although a group of survivors describe features of migraine; a focal headache, pounding in nature, and associated with photophobia, photophonia and nausea and vomiting. These headaches tended to occur almost daily or every few weeks. Others share more features with tension headaches describing band-like, constant head pains [32]. Some cases have been associated with ocular symptoms and musculoskeletal disease both of which are common, but no causal link has been identified.

Treatment is with simple analgesia and conservative measures such as rest and hydration with some improvement. WHO recommendations suggest b-blocker or amitriptyline treatment. A small group of headache patients with migrainous features have been reported to receive some benefit from b-blocker treatment however interventional study evidence is lacking.

6.10.4 Stroke

Although relatively rare, cases of stroke in both convalescent and early recovery stages following discharge are reported [31, 32]. Estimated rates of stroke in the EVD survivor population range from the recent West African outbreak are between 0.07 and 2.14% (95% CI). In keeping with the reported prothrombotic tendency of survivors, on CT scanning the territory affected is consistent with thromboembolic disease, rather than heamorrhage, and large vessel territories may be affected (see Fig. 6.10). Stroke in EVD survivors has also been associated with significant levels of physical disability and co-existing mental health disorders. In common with stroke from other aetiology, management should focus on early rehabilitation in the most suitable setting available

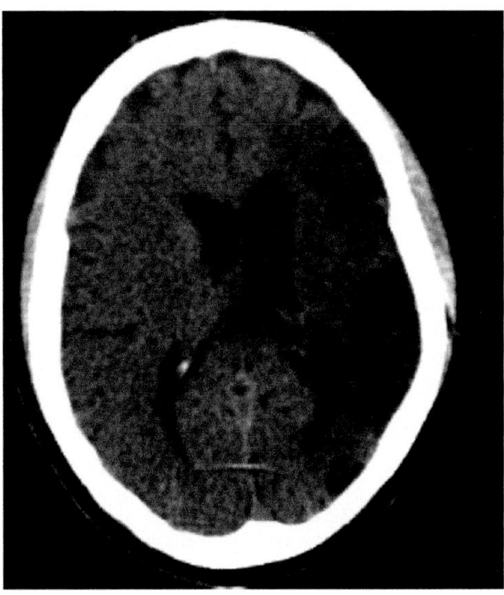

Fig. 6.10 Non-contrast brain CT scan. Male survivor, age 42. Extensive gliosis within the left MCA territory is in keeping an with old infarct with ex-vacuo dilatation of left lateral ventricle due to hemispheric volume loss

(ideally a stroke unit), adequate mental health support and secondary prevention through risk factor modification. The role for long-term antiplatelets is unclear.

6.10.5 Other Neurological Sequelae

A broad range of other neurological sequelae have been reported, the most significant of which are sensory neuropathies, plexopathies and mononeuropathies, and auditory symptoms including deafness. Other potential clinical signs include frontal lobe release signs, saccadic eye movements and gait disturbance [31, 32].

Peripheral sensory neuropathy has been reported in up to one quarter of EVD survivors, and this is characterized by an asymmetrical glove and stocking distribution with all modalities affected. Isolated peripheral nerve lesions include brachial plexus injuries [31, 32]. Management should include investigation and treatment for all other causes, complimented by rehabilitation and occupational therapy.

Although deafness is common in patients recovered from Lassa heamorrhagic fever, reports from previous epidemics were that objective findings of deafness were not more common in survivors, in comparison to controls [10, 29]. Auditory symptoms from the recent West African outbreak include tinnitus in up to 20% of patients, and hearing loss ranging from mild to profound deafness in between 2 and 7% [14, 15], with symptom onset correlated around the time of discharge from acute EVD. No characterization of the aetiology of deafness have been made.

6.11   Gastrointestinal

A significant proportion (12–35%) of survivors complain of epigastric, burning or colicky pains, not associated with nausea or vomiting or evidence of gastrointestinal bleeding in the weeks and months following discharge. Examination reveals epigastric tenderness and the pain often responds to simple antacid treatment [27, 40, 54] (Healy—personal communication).

6.12   Dermatology

In the first 30 days, 61% of survivor's report itchy skin, skin rash, infection and/or desquamation. The incidence of these symptoms fell to 43% by 90 days [7] with further improvement in the months following. Itching may or may not be associated with a rash and responds to chlorphenamine. A visible rash is often associated with dry skin, sometimes with desquamation of the hands or feet (Fig. 6.11). Appropriate treatment includes topical emollients. The aetiology of the rash is unclear, however given its reducing frequency in the early stages post discharge, and the large amounts of chlorine in treatment units, some staff and survivors attributed the rash to a reaction to chlorine vapour or solutions. Alopecia is infrequent (1%) and tends to resolve over months.

6.13   Cardiology

In eight patients managed in the US, 50% had a persistent tachycardia, present from discharge that resolved within 8 weeks [55]. A further case reported a normal echocardiograph (ECHO) and electrocardiogram (ECG) with a persistent tachycardia of 120–150, that did not respond to metoprolol or clonidine but did spontaneously resolve 2 weeks following suppression of the viraemia. Other rare early cardiological complications include clinical presentations of myocarditis and pericarditis with ECG and ECHO correlation [7, 56].

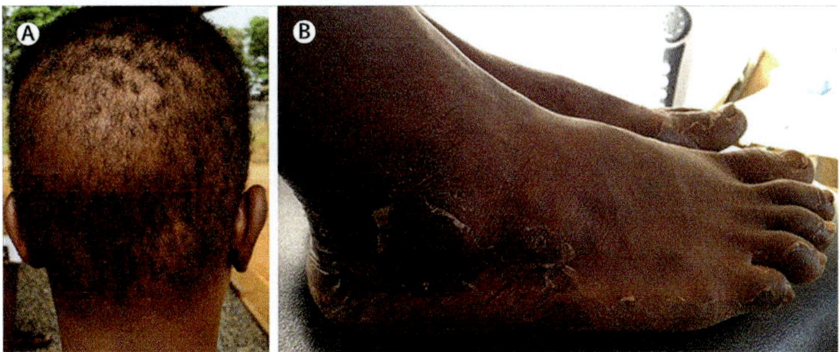

Fig. 6.11 Skin disorders in Ebola virus disease survivor. (**a**) Diffuse alopecia and (**b**) skin desquamation. Photographs by Pauline Vetter. Reprinted from The Lancet Infectious Diseases Volume 16, Issue 6, Pages e82-e91 (June 2016) DOI: 10.1016/S1473-3099(16)00077-3 with permission from Elsevier

6.14   The Post-Ebola Syndrome: A Holistic View

We are still some way from characterizing and defining a "Post-Ebola Syndrome". Whilst this may reflect that the reality of a list of non-specific symptoms, further detail from larger cohorts may provide more granularity to a definition. In addition, we are yet to understand how common co-existing conditions, such as HIV, Hepatitis B or Diabetes, and their treatments may affect the pattern and frequency of symptoms. There are some concerns, from evidence of latency of EBoV in the semen and isolated reports of EBoV viral recrudescence in HIV infected individuals, that HIV may impair immunity from EBoV although more evidence is required. For the moment, we can make the following general statements regarding the "Post Ebola Syndrome":

- Survivors are more likely to experience symptoms if they had them during their acute disease [15].
- In general adult survivors seem to experience more symptoms than children (aged less than 18 years) [5, 15].
- There is correlation between those with neurological symptoms, and ocular symptoms and musculoskeletal symptoms [15]. Mechanistically, the neurology and ocular link is supported by evidence that the distribution of Ebola retinal lesions seem to fit with neurotropic spread through the eye [46].
- Although the majority of survivors retain good function there is a small, heterogeneous subgroup who experience significant physical symptoms and co-existing mental health disorders. The design of survivor services should ensure this group is actively sought out and provided for [31, 32]

6.14.1 An Approach to the Ebola Survivor

Bearing in mind the breadth of survivor complications, Box 6.5 provides an approach to an initial clinic appointment.

Box 6.5 An Approach to the Ebola Survivor in Clinic—First Appointment

History and Examination	A full medical history with particular focus on ophthalmic, neurological and rheumatological symptoms, sexual and reproductive health and nutrition.
	Psychosocial review including sensitive questioning regarding general mood and well-being (including suicidal ideation, if appropriate), experience of stigma, current economic status, social support network and dependents, and any substance abuse.
	Examination should include as routine, neurological examination, cognitive screen and eye examination (with slit lamp if available)[a]

(continued)

Box 6.5 (continued)

Investigations	**Mandatory** Blood tests—baseline full blood count, renal and liver function Pregnancy test EBoV RT-PCR semen testing (if available) EBoV RT-PCR breast milk testing (if appropriate/available) HIV counselling and testing	**Possible other tests, as indicated from history/examination** Blood tests—TFT/CRP/ESR Joint plain X-rays Brain imaging—CT/MRI STI screening **If concern of possible EBoV relapse/febrile illness** Serum EboV RT-PCR If clinical evidence of meningitis CSF EBoV RT-PCR Testing for other causes of febrile illness—TB/Malaria/Typhoid/ Bacterial or viral infection
Principles of Management	Symptomatic treatment for main issues/concerns Nutritional advice and support Identify high symptom burden individuals, encourage utilization of support and specialist services, close follow-up in clinic Routine referral for ophthalmology review and for specialist services as appropriate Psychiatric first aid and referral to psychosocial clinic (ideally onsite of nearby) Counselling male survivors on viral persistence in semen and importance of barrier methods of contraception Counselling any pregnant female survivors on possible risks of pregnancy and importance of close follow in clinic/health centre with capability to manage EVD positive fetus/products of conception Follow-up ideally at 2 weeks, then monthly up until 6 months. Ideally, patients could transfer to local, existing health services within 8 weeks	

[a]Eye screening referral is routine, however urgent referral may be necessary

6.14.2 An Approach to the Febrile EVD Survivor

Although rare—relapse with symptoms of fever, uveitis and meningitis, with associated EBoV viraemia and isolation from the CSF has been reported up to 9 months following initial clearance from serum. As such, the assessment of the EVD survivor with fever, most particularly in the first year following discharge, should be approached with sensitivity and care. The following principles may be utilized (adapted from WHO guidance [4]):

- If a patient has fever and/or symptoms consistent with acute EVD, full IPC measures for EVD should be put in place.
- Concerning features which may suggest EVD relapse include fever with uveitis and/or meningitis.

- A serum sample should be sent for EVD RT-PCR, along with CSF samples if there is a concern of meningitis. Joint aspirate and aqueous humour aspirate may also be sent, if clinically indicated and readily available.
- Test and treat for alternative causes of a febrile illness which may be common to the region e.g. TB/Malaria.
- If index of suspicion is high, or EVD relapse is confirmed, refer to the appropriate specialist centre.

6.14.3 A Systems Approach to Ebola Survivor Care

Vertical systems are a vital part of control of an EVD outbreak. However, in the aftermath of an outbreak the transition to reframing the outbreak in the context of the health system is essential. Which elements of survivor care should remain under the care of Ebola focused systems, and when and how elements should be reintegrated into other systems and services is a debated issue.

Arguments supporting the provision of dedicated services and programmes for survivors are that survivors represent a traumatized and marginalized group with specific medical and psychosocial needs. In addition, survivors are of public health importance for the prevention of potential further outbreaks. Potential arguments supportive of early reintegration to existing services would be that an early focus of strengthening existing healthcare and public health systems is the key to prevention of future epidemics, and that reintegration to existing services, if done well, will reduce stigmatization. The ideal model is likely to be a balance of both, however potentially could reflect the following points:

- As occurred in the West Africa outbreak, Ebola Treatment units should run survivor clinics for initial consultations with the aim of providing continuity care and support for immediate and early (<8 weeks) complications.
- For complex patients, national level specialist level services with clear referral pathways for mental health, ophthalmology, rheumatology, neurology, sexual and reproductive health should be created.
- Following this initial period, attempts should be made to reintegrate care for general medical and psychosocial complications. This should be accompanied by educational resources and training for healthcare professionals, to enable them to deal with complications and refer to specialist services when necessary.
- Sexual and reproductive health should be provided free to male and female survivors for at least 18 months following the end of the epidemic. Given the risk of infection and importance for public health, free semen and breast milk testing, linked to sexual and reproductive health education, health promotion and free barrier contraception, should be provided.
- Eye screening should be provided free to survivors for at least 18 months following the end of the epidemic. Ideally, services should be integrated into national systems.
- Provided attention is given to ensuring high quality care for all survivors and integration into existing national systems, cohort studies can provide a good

model to provide quality services continuity of care for survivors, whilst also serving to advance knowledge.
- National funding should be made available for:
    - Basic needs support including food, water and housing in the immediate aftermath of the outbreak
    - Community-based programmes to tackle cross-cutting issues including stigmatization and re-integration into the community of survivors, family tracing, short-term economic support and re-establishing livelihoods and innovative and culturally relevant methods of grieving and remembering for those affected by the outbreak
    - Creation of national and local Ebola Survivor associations to empower survivors

> **Box 6.6 Integration of Psychosocial Survivor Care Health in Sierra Leone— Example (Courtesy of Dr. A Walder)**
> Prior to the EVD outbreak, the mental health service in Sierra Leone was limited; with only one psychiatrist and one psychiatric hospital located in Freetown there was very little access to care. During the outbreak there was an increased number of people reporting mental health and psychosocial problems.
>
> In a Ministry led effort to decentralize services, 21 newly trained mental health nurses were sent to the districts to create district mental health units (DMHU). The units began to provide nurse-led, integrated mental health and psychosocial support (MHPSS) to the entire population during the emergency response.
>
> Those affected by Ebola, including survivors, healthcare workers, burial team workers and affected family members, could access the service at the same point of entry as the district population. The DMHUs were developed in partnership with King's Sierra Leone Partnership and Enabling Access to Mental Health (a local non-governmental organisation, funded by CBM) who provided initial supervision and technical support. This has continued and since expanded with internationally trained Sierra Leoneon psychiatrists providing supervision and populating the referral pathways for specialist review when necessary. Between January 2015 and January 2017 327 survivors and 74 non-infected people whose mental health was directly affected by Ebola were seen in DMHUs across the country.
>
> This effort has ensured integrated mental health and psychosocial support for those affected by EVD will continue and has served to develop the mental health services throughout the country for all to access.

Any size of Ebola outbreak is a tragedy for any community and country. The race to control transmission and treat those affected is often frantic and demanding for those involved. As the response changes to focus on those left behind, it provides the opportunity to reflect on the drivers and processes which may have helped and hindered the transmission of the virus. Services put in place for survivors should serve the purpose of not just of caring for survivors but also: supporting health systems to ensure future infectious disease cases are notified and managed promptly and effectively, training more and better healthcare professions to replace those lost during the outbreak and support those left in place, and performing research and evaluation to provide more effective care for current and future EVD survivors.

6.15 Future Directions for Ebola Survivor Care and Research

There are over 10,000 survivors from the West African Ebola outbreak, and much is to be learnt over the coming years. Ongoing studies may use emerging evidence to produce high quality case-control data which may provide more clarity as to whether a "Post Ebola Syndrome" does exist. The earliest intervention study data is likely to emerge from EBoV clearance studies in the semen (e.g. GS-5734) and treatment of the uveitis. Further to this, evidence and experiences guiding the management of common complications such as arthralgia and headache would be helpful. Case studies and evaluations of survivor reintegration and stigma reduction programs, ideally with robust measurement of adjustment and mental health outcomes, would also be welcomed. For current survivor cohorts, long term outcomes, especially accounting for immunosuppressive disease (e.g. HIV) and medications, and peadiatric populations are of interest. Finally, although ring vaccination has proved successful and hopefully will limit the extent of future outbreaks, more in-depth clinical observational data of immediate and early complications of EVD would add aid future management.

References

1. Purpura LJ, Rogers E, Baller A, White S, Soka M, Choi MJ, Mahmoud N, et al. Ebola virus RNA in semen from an HIV-positive survivor of Ebola. Emerg Infect Dis. 2017;23(4):714–5. https://doi.org/10.3201/eid2304.161743.
2. Mate SE, Kugelman JR, Nyenswah TG, Ladner JT, Wiley MR, Cordier-Lassalle T, Christie A, et al. Molecular evidence of sexual transmission of Ebola virus. N Engl J Med. 2015;373 (25):2448–54. https://doi.org/10.1056/NEJMoa1509773.
3. Sissoko D, Duraffour S, Kerber R, Kolie JS, Beavogui AH, Camara AM, Graldine C, et al. Persistence and clearance of Ebola virus RNA from seminal fluid of Ebola virus disease survivors: a longitudinal analysis and modelling study. Lancet Glob Health. 2017;5(1):e80–8. https://doi.org/10.1016/S2214-109X(16)30243-1.
4. World Health Organization. Clinical care for survivors of Ebola virus disease. 2016, no, January 1–31.
5. Vetter P, Kaiser L, Schibler M, Ciglenecki I, Bausch DG. Sequelae of Ebola virus disease: the emergency within the emergency. Lancet Infect Dis. 2016a;3099(16). https://doi.org/10.1016/S1473-3099(16)00077-3 (World Health Organization. Published by Elsevier Ltd/Inc/BV. All rights reserved).
6. Howlett P, Brown C, Helderman T, Brooks T, Lisk D, Deen G, Solbrig M, Lado M. Ebola virus disease complicated by late-onset encephalitis and polyarthritis, Sierra Leone. Emerg Infect Dis. 2016;22(1):150–2. https://doi.org/10.3201/eid2201.151212.

7. Tiffany A, Vetter P, Mattia J. Ebola virus disease complications as experienced by survivors in Sierra Leone. Clin Infect. 2016. https://cid.oxfordjournals.org/content/62/11/1360.full

8. Kreuels B, Wichmann D, Emmerich P, Schmidt-Chanasit J, de Heer G, Kluge S, Sow A, et al. A case of severe Ebola virus infection complicated by gram-negative septicemia. N Engl J Med. 2014, October, 141022140021004. https://doi.org/10.1056/NEJMoa1411677.

9. Ksiazek TG, Rollin PE, Williams AJ, Bressler DS, Martin ML, Swanepoel R, Burt FJ, et al. Clinical virology of Ebola hemorrhagic fever (EHF): virus, virus antigen, and IgG and IgM antibody findings among EHF patients in Kikwit, Democratic Republic of the Congo, 1995. J Infect Dis. 1999;179(Suppl 1):S177–87. https://doi.org/10.1086/514321.

10. Rowe AK, Bertolli J, Khan AS, Mukunu R, Bressler D, Williams AJ, Peters CJ, et al. Clinical, virologic, and immunologic follow-up of convalescent Ebola hemorrhagic fever patients and their household contacts, Kikwit, Democratic Republic of the Congo. J Infect Dis. 1995;3:28–35.

11. Vetter P, Li AF, Schibler M, Jacobs M, Bausch DG, Kaiser L. Ebola virus shedding and transmission: review of current evidence. J Infect Dis. 2016b; 214(Suppl 3):177–84. https://doi.org/10.1093/infdis/jiw254.

12. McElroy AK, Akondy RS, Davis CW, Ellebedy AH, Mehta AK, Kraft CS, Marshall Lyon G, et al. Human Ebola virus infection results in substantial immune activation. Proc Natl Acad Sci USA. 2015;112(15):4719–24. https://doi.org/10.1073/pnas.1502619112.

13. Kibuuka H, Berkowitz NM, Millard M, Enama ME, Tindikahwa A, Sekiziyivu AB, Costner P, et al. Safety and immunogenicity of Ebola virus and marburg virus glycoprotein DNA vaccines assessed separately and concomitantly in healthy Ugandan adults: a phase 1b, randomised, double-blind, placebo-controlled clinical trial. Lancet. 2015a;385(9977):1545–54. https://doi.org/10.1016/S0140-6736(14)62385-0.

14. Mattia JG, Vandy MJ, Chang JC, Platt DE, Dierberg K, Bausch DG, Brooks T, et al. Early clinical sequelae of Ebola virus disease in Sierra Leone: a cross-sectional study. Lancet Infect Dis. 2016;16(3):331–8. https://doi.org/10.1016/S1473-3099(15)00489-2.

15. Etard J-F, Sow MS, Leroy S, Touré A, Taverne B, Keita AK, Msellati P, et al. Multidisciplinary assessment of post-Ebola sequelae in Guinea (Postebogui): an observational cohort study. Lancet Infect Dis. 2017;17:545–52. https://doi.org/10.1016/S1473-3099(16)30516-3.

16. Zeng X, Blancett CD, Koistinen KA, Schellhase CW, Bearss JJ, Radoshitzky SR, Honnold SP, et al. 2017. Persistent asymptomatic Ebola virus infection in Rhesus Monkeys. Nat Microbiol 17113: 1–11. doi:https://doi.org/10.1038/nmicrobiol.2017.113.

17. Carpenter A, Cox AT, Marion D, Phillips A, Ewington A. A case of a chlorine inhalation injury in an Ebola Treatment Unit. J R Army Med Corps. 2015;1–3. https://doi.org/10.1136/jramc-2015-000501. (Epub ahead of print).

18. Rosenke K, Adjemian J, Munster VJ, Marzi A, Falzarano D, Onyango CO, Ochieng M, et al. Plasmodium parasitemia associated with increased survival in Ebola virus–infected patients. Clin Infect Dis. 2016;63(8):1026–33. https://doi.org/10.1093/CID/CIW452.

19. Who Meeting on Survivors of Ebola Virus Disease: Clinical Care of Survivors Meeting Report. 2015, no. August: 3–4.

20. Deen GF, Knust B, Broutet N, Sesay FR, Formenty P, Ross C, Thorson AE, et al. Ebola RNA persistence in Semen of Ebola virus disease survivors—preliminary report. N Engl J Med. 2015;151014140118009. https://doi.org/10.1056/NEJMoa1511410.

21. Sterk E. Filovirus haemorrhagic fever. Filovirus haemorrhagic fever guideline 2008; 2008.

22. Kibuuka H, Berkowitz NM, Millard M, Enama ME, Tindikahwa A, Sekiziyivu AB, Costner P, et al. The merits of malaria diagnostics during an Ebola virus disease outbreak. Emerg Infect Dis. 2015b;22(2):2448–54. https://doi.org/10.3201/eid2202.151656.

23. Kerber R, Krumkamp R, DIallo B, Jaeger A, Rudolf M, Lanini S, Bore JA, et al. Analysis of diagnostic findings from the European mobile laboratory in Guéckédou, Guinea, March 2014 through March 2015. J Infect Dis. 2016;214:S250–7. https://doi.org/10.1093/infdis/jiw269.

24. Waxman M, Aluisio AR, Rege S, Levine AC. Characteristics and survival of patients with Ebola virus infection, malaria, or both in Sierra Leone: a retrospective cohort study. Lancet Infect Dis. 2017;17(6):654–60. https://doi.org/10.1016/S1473-3099(17)30112-3.
25. Uyeki TM, Mehta AK, Davey RT Jr, Liddell AM, Wolf T, Vetter P, Schmiedel S, et al. Clinical management of Ebola virus disease in the United States and Europe. N Engl J Med. 2016;374 (7):636–46. https://doi.org/10.1056/NEJMoa1504874.
26. Calain P, Bwaka MA, Colebunders R, De Roo A, Guimard Y, Katwiki KR, Kibadi K, et al. Ebola hemorrhagic fever in Kikwit, Democratic Republic of the Congo : clinical observations in 103 patients. J infect Dis. 1999;179(Suppl 1):1–7.
27. Qureshi AI, Chughtai M, Loua TO, Kolie JP, Camara HFS, Ishfaq MF, N'Dour CT, Beavogui K. Study of Ebola virus disease survivors in Guinea. Clin Infect Dis. 2015;61(7):1035–42. https://doi.org/10.1093/cid/civ453.
28. Lötsch F, Schnyder J, Goorhuis A, Grobusch MP. Neuropsychological long-term sequelae of Ebola virus disease survivors – a systematic review. Travel Med Infect Dis. 2017;18:18–23. https://doi.org/10.1016/j.tmaid.2017.05.001.
29. Clark DV, Kibuuka H, Millard M, Wakabi S, Lukwago L, Taylor A, Eller MA, et al. Long-term sequelae after Ebola virus disease in Bundibugyo, Uganda: a retrospective cohort study. Lancet Infect Dis. 2015;15(8):905–12. https://doi.org/10.1016/S1473-3099(15)70152-0.
30. Jagadesh S, Sevalie S, Fatoma R, Sesay F, Sahr F, Faragher B, Semple MG, Fletcher TE, Weigel R, Scott JT. Disability among ebola survivors and their close contacts in Sierra Leone: a retrospective case-controlled cohort study. Clin Infect Dis. 2018;66(1):131–3. https://doi.org/ 10.1093/cid/cix705.
31. Bowen L, Smith B, Steinbach S, Billioux B, Summers A, Azodi S, Ohayon J, Schindler M, Nath A. Survivors of Ebola virus disease have persistent neurologic deficits. n.d.
32. Howlett PJ, Walder AR, Lisk DR, Fitzgerald F, Sevalie S, Lado M, et al. Case series of severe neurologic sequelae of Ebola virus disease during epidemic, Sierra Leone. Emerg Infect Dis. 2018;24(8):1412–21. https://doi.org/10.3201/eid2408.171367.
33. Dhillon P, McCarthy S, Gibbs M. Surviving stroke in an Ebola treatment centre. BMJ Case Rep. 2015;3–4. https://doi.org/10.1136/bcr-2015-211062.
34. Wilson AJ, Martin DS, Maddox V, Rattenbury S, Bland D, Bhagani S, Cropley I, et al. Thromboelastography in the management of coagulopathy associated with Ebola virus disease. Clin Infect Dis. 2016;62(5):610–2. https://doi.org/10.1093/cid/civ977.
35. Bower H, Smout E, Bangura MS, Kamara O, Turay C, Johnson S, Oza S, Checchi F, Glynn JR. Deaths, late deaths, and role of infecting dose in Ebola virus disease in Sierra Leone: retrospective cohort study. BMJ. 2016;i2403. https://doi.org/10.1136/bmj.i2403
36. De Roo A, Ado B, Rose B, Guimard Y, Fonck K, Colebunders R. Survey among survivors of the 1995 Ebola epidemic in Kikwit, Democratic Republic of Congo: their feelings and experiences. Trop Med Int Health. 1998;3(11):883–5. https://doi.org/10.1046/j.1365-3156. 1998.00322.x.
37. Rabelo I, Lee V, Fallah MP, Massaquoi M, Evlampidou I, Crestani R, Decroo T, Van den Bergh R, Severy N. Psychological distress among Ebola survivors discharged from an Ebola treatment unit in Monrovia, Liberia – a qualitative study. Front Public Health. 2016;4:142. https://doi.org/10.3389/fpubh.2016.00142.
38. Gidado S, Oladimeji AM, Roberts AA, Nguku P, Nwangwu IG, Waziri NE, Shuaib F, et al. Public knowledge, perception and source of information on Ebola virus disease – Lagos, Nigeria; September, 2014. PLoS Curr. 2015;7 (OUTBREAKS). Public Library of Science. https://doi.org/10.1371/currents.outbreaks.0b805cac244d700a47d6a3713ef2d6db.
39. Nakiyingi L, Nankabirwa H, Lamorde M. Tuberculosis diagnosis in resource-limited settings: clinical use of GeneXpert in the diagnosis of smear-negative PTB: a case report. Afr Health Sci. 2013;13(2):522–4. https://doi.org/10.4314/ahs.v13i2.46.
40. Nanyonga M, Saidu J, Ramsay A, Shindo N, Bausch DG. Sequelae of Ebola virus disease, Kenema District, Sierra Leone. Clin Infect Dis. 2016;62(1):125–6. https://doi.org/10.1093/cid/ civ795.

41. Keita MM, Taverne B, Savané SS, March L, Doukoure M, Sow MS, Touré A, Etard JF, Barry M, Delaporte E. Depressive symptoms among survivors of Ebola virus disease in Conakry (Guinea): preliminary results of the PostEboGui Cohort. BMC Psychiatry. 2017;17 (1):127. https://doi.org/10.1186/s12888-017-1280-8.
42. Gillespie DK. Psychological first aid. 1963;33(9):391–95. https://doi.org/10.1111/j.1746-1561. 1963.tb00427.x.
43. Valle C. Accepting and supporting Ebola survivors, orphans and families of Ebola patients in the community a toolkit for social mobilizers and communicators. 2014. http://another-option. com/wp-content/uploads/2014/07/AO_EBOLA_Stigma_Toolkit_Final_4.pdf
44. Kibadi K, Mupapa K, Kuvula K, Massamba M, Ndaberey D, Bwaka MA, De Roo A, Colebunders R. Late ophthalmologic manifestations in survivors of the 1995 Ebola virus epidemic in Kikwit, Democratic Republic of the Congo. J Infect Dis. 2000;179(Suppl 1):13–4.
45. Varkey JB, Shantha JG, Crozier I, Kraft CS, Marshall Lyon G, Mehta AK, Kumar G, et al. Persistence of Ebola virus in ocular fluid during convalescence. N Engl J Med. 2015;372 (25):2423–7. https://doi.org/10.1056/NEJMoa1500306.
46. Steptoe PJ, Scott JT, Baxter JM, Parkes CK, Dwivedi R, Czanner G, Vandy MJ, et al. Novel retinal lesion in Ebola survivors, Sierra Leone, 2016. Emerg Infect Dis J. 2017;23(7):1102–9. https://doi.org/10.3201/eid2307.161608.
47. Baggi FM, Taybi A, Kurth A, Van Herp M, Di Caro A, Wölfel R, Günther S, Decroo T, Declerck H, Jonckheere S. Management of pregnant women infected with ebola virus in a treatment centre in Guinea. June 2014. Euro Surveill. 2014;19.
48. Caluwaerts S, Fautsch T, Lagrou D, Moreau M, Camara AM, Gnther S, Di Caro A, et al. Dilemmas in managing pregnant women with Ebola: 2 case reports. Clin Infect Dis. 2016;62 (7):903–5. https://doi.org/10.1093/cid/civ1024.
49. Fallah MP, Skrip LA, Dahn BT, Nyenswah TG, Flumo H, Glayweon M, Lorseh TL, Kaler SG, Higgs ES, Galvani AP. Pregnancy outcomes in Liberian women who conceived after recovery from Ebola virus disease. Lancet Glob Health. 2016;4(10):e678–9. https://doi.org/10.1016/ S2214-109X(16)30147-4. (This is an Open Access article under the CC BY license)
50. Nicastri E, Balestra P, Ricottini M, Petrosillo N, Di Caro A, Capobianchi MR, Giancola ML, Ippolito G. Temporary neurocognitive impairment with Ebola virus: Table 1. J Neurol Neurosurg Psychiatry. 2016;0(0):jnnp-2016-313695. https://doi.org/10.1136/jnnp-2016-313695.
51. Chertow D, Nath A, Suffredini A, Danner R. Severe meningoencephalitis in a case of Ebola virus disease: a case report. Ann Intern Med. 2016;165(4):301–4.
52. Jacobs M, Rodger A, Bell DJ, Bhagani S, Cropley I, Filipe A, Gifford RJ, et al. Late Ebola virus relapse causing Meningoencephalitis: a case report. Lancet (Lond). 2016;6736(16):1–6. https:// doi.org/10.1016/S0140-6736(16)30386-5.
53. Billioux BJ. Neurological complications of Ebola virus infection. Neurotherapeutics. 2016;13 (3):461–70. https://doi.org/10.1007/s13311-016-0457-z.
54. Wendo C. Caring for the survivors of Uganda's Ebola epidemic one year on TT. Lancet. 2001;358(9290):1350. https://doi.org/10.1016/S0140-6736(01)06467-4.
55. Epstein L, Wong K, Kallen KJ, Ukeyi TM. Post-Ebola signs and symptoms in U.S. survivors. N Engl J Med. 2015;373(25):2484. https://doi.org/10.1056/NEJMc1511801.
56. Kibadi K, Mupapa K, Kuvula K, Massamba M, Ndaberey D, Muyembe-Tamfum JJ, Bwaka MA, De Roo A, Colebunders R. Late ophthalmologic manifestations in survivors of the 1995 Ebola virus epidemic in Kikwit, Democratic Republic of the Congo. J Infect Dis. 1999;179 (Suppl 1):S13–4. https://doi.org/10.1086/514288.

Index